THE CHEROKEE STRIP LIVE STOCK ASSOCIATION

THE CHEROKEE STRIP LIVE STOCK ASSOCIATION

Federal Regulation and the Cattleman's Last Frontier • William W. Savage, Jr.

Permission has been granted for the use of the illustration on page 81 by the National Archives, for those on pages 115 and 137 by the University of Oklahoma Press. Permission for the use of all other illustrations has been granted by the Western History Collections of the University of Oklahoma Library, Norman.

University of Missouri Press, Columbia, Missouri 65201
Printed and bound in the United States of America
Library of Congress Catalog Number 72–93932

ISBN 0–8262–0144–X

To My Parents

PREFACE

In the decade following the Civil War, interrelated factors—pacification of the Indians, construction of railroads, and decimation of the buffalo—opened the vast grasslands of the Great Plains to ever-growing herds of domestic cattle. Ranges in Texas, largely untouched by the war, became livestock breeding grounds. Cattle worth only about five dollars per head in Texas could bring between thirty-five and forty dollars per head in the East, and in 1866, Texas stockmen, seeking to capitalize on the market, began trailing their herds to railheads in Missouri and Kansas. The cattleman's domain spread north from the Rio Grande to the Canadian border; soon, grass became scarce. It was perhaps inevitable that ranchers should turn their attention to the virgin ranges of Indian Territory. The region was to become the cattleman's last frontier.

On March 7, 1883, the Cherokee Strip Live Stock Association was incorporated under the laws of the State of Kansas. Four months later, its directors leased from the Cherokee Nation over six million acres of the Cherokee Outlet, an area in northwestern Indian Territory once granted to the tribe as a perpetual corridor to the western boundary of the United States. Members of the association grazed their cattle in the Outlet until the federal government purchased it in 1890. Thereafter, the association endured as an organization in legal limbo for another three years, succumbing corporately at about the same time that the Outlet was opened to settlement.

It is not from its relationship to the Western range cattle industry that the Cherokee Strip Live Stock Association derives its historical significance. As a rule, cattlemen's organizations—roughly comparable to Eastern trade associations—were formed to regulate round-ups, keep a record of members' brands, determine ownership of mavericks, and the like. They gave substance to what had once been cow custom, or informal practice, on open, or unfenced, range. The Cherokee Strip Live Stock Association differed from the other ranchers' organizations because it was forced into existence by the regulating activities of the federal government. The laissez-faire philosophy that characterized the national government's attitude toward entrepreneurs in the rapidly industrializing East did not extend to the businessmen in the West, as evidenced by the experience of the Kansas ranchers who formed the association. Distinctly, government was not an arm of business in the West.

The government's eventual interference in the business of ranchers in the Outlet was in large measure the result of its desire to retain complete control of Indian affairs and of its view of the West as little more than potential farmland. It found an occasional ally in the settler who hoped to homestead Indian land and a constant opponent in the Cherokee who saw his tribal sovereignty threatened. Herein lies the historical importance of the association. Inquiry into its operations provides a vehicle for examining the interrelationships between four disparate elements—homesteaders, cattlemen, Indians, and the federal government—that sought to control vast areas of land in the post-Civil War West. Study of the Cherokee Strip Live Stock Association yields useful insights into the federal government's Indian and land policies and the interaction of governmental agencies that implemented them. It also constitutes a significant comment on the role of bureaucracy—the hierarchy of administrators, civil servants, and agencies comprising government in its institutional form—in the development of the West. It confirms the relatively recent and—one would think—obvious view of the cattleman as an entrepreneur, but more importantly, it provides valuable and hitherto neglected perspectives on the Indian as a businessman and the larger question of tribal involvement in the national economy. Finally, study of the association and its operations adds a new dimension to the myth of frontier violence. The entire history of ranching on the Cherokee Outlet was bloodless, and it therefore contradicts the stereotype of sanguinary showdowns between homesteaders and cowmen over quarter-sections of rangeland; in that corner of the West, at least, the violence was rhetorical.

Because of the crucial role the Cherokee Nation played in the history of the Cherokee Strip Live Stock Association, some understanding of the structure of its government will be essential to the reader. The tribal constitution, promulgated in 1839 and amended in 1866, provided for government by an elected principal chief, a bicameral legislature, and a judiciary. The legislature, or National Council, consisted of an upper house, known after 1866 as the senate, and a lower house, designated as the council. Appropriations bills originated in the senate but could be amended or rejected by the lower house. Any other legislation could originate in either house, subject to the approval or rejection of the other. The veto power of the principal chief might be overridden by two-thirds votes in both houses. In this study, the terms ***council*** and ***National Council*** will be used interchangeably; the terms ***senate*** and ***lower house*** will be used only when these distinctions are necessary.

A word about semantics is perhaps in order. In the literature of the Western range cattle industry, the designations ***Cherokee Strip*** and ***Cherokee***

Outlet are frequently applied interchangeably. The Outlet's cattlemen, as the name of their association indicates, followed that usage. The Cherokee Strip, however, was a three-mile-wide ribbon of land along the southern border of Kansas, running from the Missouri line to the 100th meridian. It resulted from an error in an early boundary survey and was ceded by the Cherokee Nation to the federal government in 1866. Strip and Outlet were contiguous but otherwise unrelated. This study will observe the proper geographical distinctions except in quoting contemporary sources that use the terms alternately.

I am endebted to a number of persons for assistance in completing this study. I extend special thanks to Miss Opal Carr, formerly of the University of Oklahoma Library; Mrs. Alice Timmons and Mr. Jack Haley, Western History Collections, University of Oklahoma Library; Mrs. Rella Looney, Oklahoma Historical Society; and Mr. Richard S. Maxwell and Mr. Milton Ream, National Archives, all of whom greatly facilitated my investigations. A grant from the Graduate Studies Committee of the Department of History, University of Oklahoma, financed much of the research. Donald J. Berthrong of Purdue University, Gilbert C. Fite of Eastern Illinois University, and Savoie Lottinville and Norman Crockett of the University of Oklahoma offered many valuable criticisms and suggestions. My greatest intellectual debt, however, is to Arrell M. Gibson of the University of Oklahoma, a wise and patient scholar whose confidence and encouragement have been an inspiration. I am grateful also to Miss Alexis Rodgers, who typed the manuscript. And finally, to my wife Sue, who nobly tolerated my frequent and prolonged excursions into the nineteenth century and did much to expedite my work, I express my deepest gratitude.

W. W. S.
Norman, Oklahoma
February, 1973

CONTENTS

THE CHEROKEE STRIP LIVE STOCK ASSOCIATION

Cherokee Lands in Indian Territory, 1866–1893

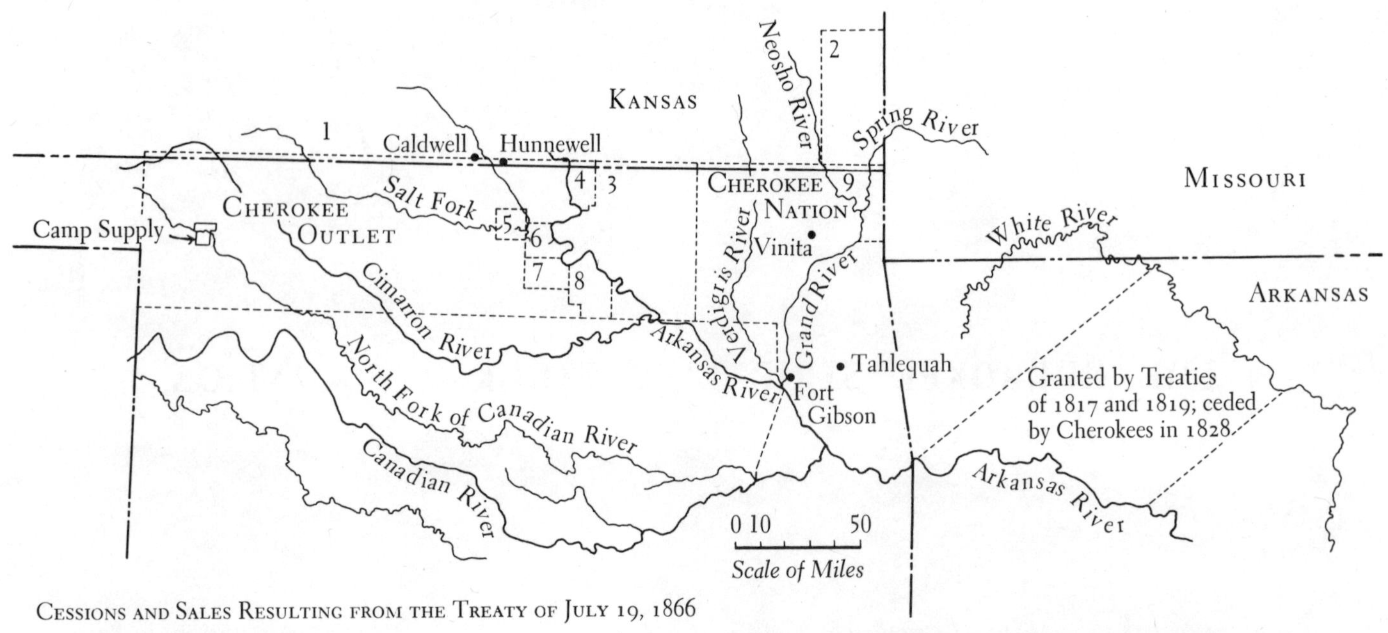

Cessions and Sales Resulting from the Treaty of July 19, 1866

1. "Cherokee Strip," ceded 1866
2. "Neutral Lands," ceded 1866
3. Sold to Osages
4. Sold to Kaws
5. Sold to Nez Percés; occupied by Tonkawas after 1885
6. Sold to Poncas
7. Sold to Otoes and Missouris
8. Sold to Pawnees
9. Peorias, Quapaws, Ottawas, Shawnees, Modocs, Wyandottes, and Senecas, 1832–1873

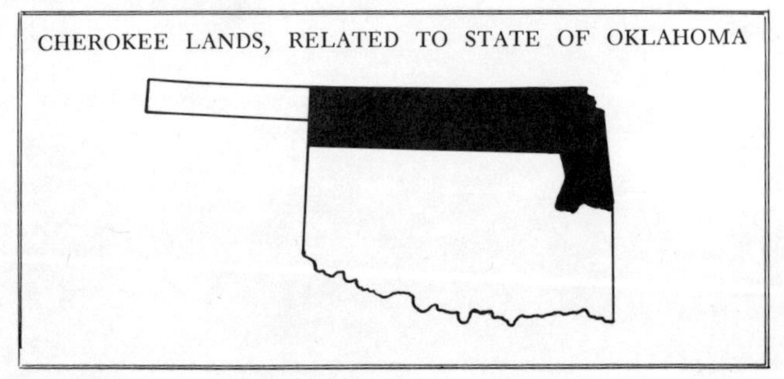

I. THE LAND

The Western range cattle industry rose like a phoenix from the ashes of the Civil War. Texas longhorns, left to graze untended in regions far removed from the holocaust, multiplied rapidly during the war years. Beef was so plentiful as to be virtually worthless on local markets, and stockmen who returned to the state in 1865 discovered that they were cattle-poor. But they soon learned that vast riches lay elsewhere. The East, on the verge of industrial expansion, needed food. There, demand exceeded supply, and beef cattle frequently brought as much as forty dollars per head at auction. Texas ranchers, eager to profit from the windfall prices and stimulated by news of Joseph G. McCoy's market at Abilene, Kansas, started the long trail drive to railheads in Missouri and Kansas. The northward migrations of longhorns, beginning in 1866, marked the initial expansion of the business that was to dominate the economic life of the Great Plains for a quarter of a century.[1]

In the process of breaking trails, first to the rail towns and later to the feeding grounds of Wyoming, Montana, and the Dakotas, cattlemen crossed the lush grasslands of northern Indian Territory. The Chisholm Trail, first and most famous of the cattle routes, bisected a fertile expanse of land labeled on maps as the Cherokee Outlet. Its remoteness from Indian settlements made it an attractive

1. The origins of the postwar cattle business are discussed in Edward Everett Dale, *The Range Cattle Industry: Ranching on the Great Plains from 1865 to 1925*; Walter Prescott Webb, *The Great Plains*, chap. 6; and Wayne Gard, *The Chisholm Trail*, chaps. 3, 4, and 5. Useful too is Joseph G. McCoy, *Historic Sketches of the the Cattle Trade of the West and Southwest*, the first published history of the business.

grazing area, and stockmen moving north from Texas found suitable winter pasture there. As the cattle kingdom grew, grassland became scarce, and ranchers in western Kansas often drifted their herds south to the Outlet whenever they ran short of pasture. These stockmen met little competition from native herdsmen because the territory had been "nearly denuded" of cattle by California-bound buyers in the late 1850s and by foraging parties during the Civil War. A movement that began perhaps free of intent to trespass became full-scale exploitation by the late 1870s. Confrontation with the Outlet's rightful owner, the Cherokee Nation, was not long in coming.[2]

2. Grant Foreman, *The Five Civilized Tribes*, p. 81; and Arrell M. Gibson, *Oklahoma: A History of Five Centuries*, p. 207.

In 1817 the federal government implemented a policy of Indian removal that was at first conducted on a voluntary basis. Cherokees who agreed to leave their homes in the Southeast received compensatory grants of land in Arkansas. In 1828 these Western Cherokees were encouraged to emigrate to new lands in the northeastern corner of Indian Territory. To prevent their future encirclement by whites, they were assured by treaty of a perpetual corridor to the western limits of the United States.

The Cherokees wished to have access to the hunting grounds beyond the 100th meridian; the idea of a perpetual outlet to the West seems to have been originated by Thomas Jefferson as early as 1808. Those who interpret **civilized,** in the sense that the term was applied to the five major tribes of the southeastern United States, to mean ***sedentary*** may have difficulty in believing that Cherokees were concerned about such matters as the availability of game. Yet, at the end of the eighteenth century, some Cherokees migrated to the White River country of northern Arkansas for the stated purpose of being able to hunt on their own land. At least one eyewitness, the Count de Pourtalès, who accompanied Washington Irving on his western junket in 1832, observed a Cherokee hunting party en route to the Salt Plains in the north-central portion of the Outlet. This early provision of hunting grounds was the origin of the Cherokee Outlet, a belt of land sixty miles wide, running from the 96th meridian to the 100th meridian, the boundary established by the Adams-Onis Treaty of 1819.[3]

The right of access to the West via the Outlet was reconfirmed in 1835 by the Treaty of New Echota, which provided for the removal of the Eastern Cherokees. The fact that since 1855 only a few Cherokees used the Outlet for hunting purposes was evidently sufficient reason to find other uses for the land. The extent of the Outlet, originally comprising nearly 6,574,486 acres, was reduced in 1866 by a treaty that allowed the federal government to purchase a part of the land for the settlement of friendly tribes. The Osages obtained most of the land lying between the 96th meridian and the Arkansas River; the Kaws received a small portion in the northwestern corner; the Pawnees, Poncas, Nez Percés, Otoes, and Missouris were settled along the west bank of the Arkansas and on the Salt Fork. The Nez Percés left the Outlet in 1885, and their land was taken by the Tonkawas. During this time of resettlement, the Army established Camp Supply on the North Fork of the Canadian River in the western part of the Outlet in 1868. Thus, when the cattlemen arrived, the Outlet consisted of about 6,022,754 acres of land.[4]

The Cherokee Outlet was well suited to sustain the herds of cattle that moved into it in the 1870s. It was relatively flat country, broken

3. Government thinking on the subject is outlined in U. S., Congress, House, 51st Cong., 2d sess., 1890, Report 3768, pp. 6–7. The most complete account is Berlin B. Chapman, "How the Cherokees Acquired the Outlet," *The Chronicles of Oklahoma*, 15:1 (March 1937), 30–49. That the outlet idea was from the first closely related to the problems of Cherokee removal is indicated by both Chapman, ibid., and Grant Foreman, *Indians and Pioneers: The Story of the American Southwest Before 1830*, p. 31. For the Cherokees' hunting practices, see Foreman, p. 26. *On the Western Tour with Washington Irving: The Journal and Letters of Count de Pourtalès*, p. 58.

4. The details of the Cherokees' removal and the reunification of the tribe are discussed in Grant Foreman, *Indian Removal: The Emigration of the Five Civilized Tribes of Indians*, chaps. 18–24. House Report 3768, op. cit., p. 17; for the acreage settled by friendly tribes, see p. 1. For the precise location of tribes on the Outlet, see John W. Morris and Edwin C. McReynolds, *Historical Atlas of Oklahoma*, Map 20. For details on the creation of Camp Supply, see Robert C. Carriker, *Fort Supply, Indian Territory: Frontier Outpost on the Plains*.

occasionally by low, rolling hills and watered by four major streams—the Arkansas, the Salt Fork, the Cimarron, and the North Canadian. The basis of cattle country, however, is grass, and in this respect the Outlet had few equals. Its natural vegetation depended on rainfall, which was usually ample, varying annually from twenty-two inches in the westernmost portion to thirty-two inches along the Arkansas River. Tall, sod-forming grasses like big bluestem, Indian grass, and switchgrass were indigenous to the eastern region, while short, sod-forming buffalo grass and bunch grass like blue grama and sideoats grama were characteristic in the central and western areas. Occasionally, tall grasses like giant wild rye were found farther west, where there was wet, saline soil. The distribution of these grasses in relation to soil conditions and climate made the Outlet an excellent pasture. This potentiality had not escaped the notice of early travelers through the region. One of them, Capt. Nathan Boone, who led an expedition from Fort Gibson to the Great Salt Plains in the north-central portion of the Outlet during the summer of 1843, commented on the "luxuriant" vegetation just beyond the Arkansas River and the "fine grass" along the Cimarron River and the North Fork of the Canadian. Twenty-five years later, Texas cattlemen were to make that discovery for themselves.[5]

If Texas ranchers after the Civil War were cattle-poor, it is not inaccurate to say that their future partners, the Cherokees, were land-poor to an equal degree. The tribe held the Outlet in fee simple but could derive no revenue from it. Long since, the United States boundary had pushed to the Pacific and the last Cherokee hunting party had roamed west of the Arkansas. Moreover, the Treaty of 1866

5. For landforms, rainfall distribution, and rivers, see Morris and McReynolds, *Historical Atlas*, Maps 3, 4, and 5. Sellers G. Archer and Clarence E. Bunch, *The American Grass Book: A Manual of Pasture and Range Practices*, pp. 127, 194 ff. "Captain Boone's Journal of an Expedition over the Western Prairies," in Louis Pelzer, *Marches of the Dragoons in the Mississippi Valley*, pp. 193, 225. Julian Fessler, ed., "Captain Nathan Boone's Journal," *The Chronicles of Oklahoma*, 7:1 (March 1929), 58–105, contains a detailed map of Boone's route.

had effectively severed the Outlet from the rest of Cherokee domain. Rarely did Cherokee stockmen venture into the Outlet, and then only with a few head of cattle. Despite its excellence as grazing land, the Outlet was simply too far from their farmland for Cherokees to use it profitably. Nor did the tribe own sufficient livestock to occupy more than a fraction of the available pasture. It was an untenable situation, and so the Cherokee Nation sought a remedy in the treaty that had caused the problem. If the federal government could purchase portions of the Outlet for purposes of relocating friendly tribes, then, the Cherokees' leaders reasoned, it could just as easily purchase the entire six million acres for whatever purposes it saw fit. Consequently, the Cherokee National Council in 1872 began instructing its delegates to Washington to negotiate for the sale of the Outlet.[6]

Fortunately for both the Cherokees and cattlemen, Congress was reluctant to appropriate the funds necessary to buy the Outlet. The Indians were not blind to the horned invasion that was taking place on their land beyond the Arkansas, but neither were they quick to see its advantages for them. They did levy a tax, which in 1867 amounted to ten cents per head, on livestock passing through the Outlet, and although this amount was increased in 1869, ten years elapsed before the Nation imposed a grazing tax.[7]

Although the United States Senate Committee on Indian Affairs approved the idea of a tax on stockmen grazing herds on Cherokee land in 1870 and the Department of the Interior endorsed the committee's decision in 1872, not until 1878, when the Senate Judiciary Committee confirmed the Cherokees' right of taxation, did the Indians attempt to collect grazing fees. In 1879 the Nation appointed L. B.

6. Extracts of instructions relating to the Cherokee Outlet, 1873–1888, are contained in House Report 3768, op. cit., pp. 22–25.

7. U. S., Congress, Senate, 41st Cong., 2d sess., 1870, Report 225, p. 1.

Bell as Special Tax Collector. By October, he and his assistants had brought $1,100 into the coffers of the Cherokee Treasury. Bell's presence on the Outlet range antagonized cattlemen who had hitherto paid no tax, and although they protested imposition of the tax, the first criticism of the Cherokees' action came from another quarter.[8]

On August 4, 1879, W. H. Whiteman, agent of the Ponca and Nez Percé, complained about Cherokee taxation to the Commissioner of Indian Affairs. He contended that Bell had abused his authority by granting grazing licenses—permits given to taxpayers—to some undesirable characters. On August 28, because of the furor that Whiteman's charges created, Cherokee Principal Chief Charles Thompson appointed William McCracken and Elias C. Boudinot to investigate conditions beyond the Arkansas. Their report, which exonerated Bell, provided a detailed picture of early Outlet ranching operations.[9]

There were, according to McCracken and Boudinot, approximately twenty-five taxpaying stockmen grazing 20,000 head of cattle on the Outlet in 1879; thirty others had paid no taxes. Together, these cattlemen employed between 180 and 200 cowboys, and each rancher had made improvements on his pasture. Probably these were crude structures, built to serve as headquarters. The investigators noted that they "were only temporary and could easily be abandoned" when ranchers left the range. They appended to their report statements from the military commander at Arkansas City and from an agent for the "Pan Handle Texas Stage route," both testifying as to the impermanency of cattlemen's improvements.

Boudinot and McCracken heard many complaints of exorbitant taxation. Cattlemen said that they would not object to reasonable

8. Letter of William P. Adair and Dan'l H. Ross, Cherokee Delegation, Giving to the Cherokee People an Account of Their Mission to Washington, July 8, 1878. Printed copy in Cherokee Nation Papers, Western History Collections, University of Oklahoma Library, Norman. Papers hereafter cited as CNP, Collections as WHC. *The Cherokee Advocate* (Tahlequah), February 6, 1885.

9. William McCracken and E. C. Boudinot, Jr., to Charles Thompson, October 23, 1879, File: Cherokee—Strip (Tahlequah), 1879, Indian Archives Division, Oklahoma Historical Society, Oklahoma City. Archives hereafter cited as IAD, Society as OHS.

Charles Thompson, Principal Chief
of Cherokee Nation in 1879

Dennis Wolfe Bushyhead, Principal Chief of Cherokee Nation, 1879–1887

rates but currently they were being asked to pay amounts approaching 5 per cent of their original investment. Some taxpayers reported that assessments were not the same for all stockmen, but the investigators told the Principal Chief that "Fifty cents per head per annum was the tax rate collected by L. B. Bell." Boudinot and McCracken concluded their presentation by suggesting that the National Council consider "proper legislation" for the "proper management" of the Outlet, because, as they said, "We think that a very large revenue could be secured to the Cherokee Nation that is and has been justly due." [10]

In 1879 Dennis Wolfe Bushyhead succeeded Charles Thompson as Cherokee Principal Chief, and it became his responsibility to assess Boudinot's and McCracken's report. On November 19 he sent his recommendations to the National Council, then in session in Tahlequah, the Cherokee capital. Bushyhead prefaced his remarks by referring to Article XVI of the Treaty of 1866, which reads in part, "The Cherokee nation to retain the right of possession of and jurisdiction over all of said country west of 96° of longitude until . . . sold and occupied." His purpose was "to show that the ***right of possession*** and . . . the profits of possession in the lands west of the Arkansas River remain in the Cherokee Nation until the . . . lands are settled and sold." He acknowledged that Bell and his deputies had been in the main unsuccessful in collecting the grazing tax, and he blamed their failure on "the excessively high tax rates." Bushyhead proposed that assessments be fixed at fifty cents per head for a six- or eight-month period. The tax could then be decreased proportionately for longer terms of occupancy and increased for shorter ones. He predicted that within one year "the revenue from such source alone,

10. McCracken and Boudinot, op. cit. Statement of Lt. Cushman and Statement of H. A. Todd, undated, File: Cherokee—Strip (Tahlequah), 1879, IAD, OHS. McCracken and Boudinot.

D. W. Lipe, Treasurer, Cherokee Nation

to the Nation, would amount to at least twenty thousand . . . Dollars." He concluded,

> I see no good reason why the profits of millions of acres of fine pasturage yet remaining to us west of Arkansas River should not be shared in proper proportion between those who have used, and will use, those lands to the extent represented, without any right to use them unpermitted, and those who have the right alone, at the present time to use them and to grant others permission to do so—i.e., the Cherokee Nation.[11]

Cherokee Treasurer D. W. Lipe personally supervised tax collection in the Outlet in the spring of 1880. He met the same resistance that Bell had encountered in 1879. Additional meetings with cattlemen in Caldwell, Kansas, headquarters for many of those who grazed stock in Indian Territory, did little to further his work. Lipe reported to Bushyhead in June that, regardless of whatever action the National Council might take to set tax rates, cattlemen would not consent to pay more than twenty-five cents per head annually. The Treasurer had offered monthly rates of five cents per head "on any receipt for less than a year" and a flat assessment of fifty cents per head on an annual basis. Most of the cattlemen with larger herds appeared willing to pay, Lipe said, but he reported that he received "only . . . about $1,600.00" on that trip to Caldwell. In all, Lipe managed to collect $7,620 during 1880, a figure far below that which Bushyhead had anticipated. The negotiations between the Cherokees and the cattlemen seemed to have reached an impasse; the problems over collections dragged on into the next year.[12]

In the spring of 1881 Lipe returned to Caldwell. He attended a

11. D. W. Bushyhead to the Honorable, The Senate and Council, November 19, 1879 (copy), File: Cherokee—Strip (Tahlequah), 1879, IAD, OHS. Charles J. Kappler, comp. and ed., *Indian Affairs: Laws and Treaties*, II, 947.

12. D. W. Lipe to Hon. D. W. Bushyhead, June 4, 1880, File: Cherokee—Strip (Tahlequah), 1880–1881, IAD, OHS. *The Cherokee Advocate* (Tahlequah), February 6, 1885.

ranchers' meeting in March "to look after the revenue interest of the Nation, and to obtain a general expression of their feelings in regard to paying of taxes the coming season." The cattlemen were in a bargaining mood and agreed to pay the grazing tax if the Cherokees would sanction the creation of a quarantine line between the Kansas boundary and the Nez Percé reservation for herds moving to market across Indian Territory. Lipe, eager to get on with business concerning taxes, promised to present their request to Bushyhead. The quarantine ground was eventually established, but at a point farther west, where the Chisholm Trail intersected the Kansas border. Yet, despite cattlemen's assurances, tax collection became no easier.[13]

Because grazing in the Outlet loomed large in the Cherokee Nation's fiscal plans, Lipe opened a branch office in Caldwell. In the summer of 1881 the office was staffed by J. G. Schrimsher and a deputy tax collector named Brewer. Schrimsher, a close friend of Bushyhead, ran the office with a firm hand. Upon learning that cattlemen were waiting to see whether or not the Cherokee Nation planned to enforce its revenue laws, he promptly notified delinquent taxpayers to meet their obligations by July 5 or face eviction from the Outlet. Two weeks before the deadline, he urged the Principal Chief to request federal assistance in carrying out the expulsions, should his ultimatum fail to produce the desired result. Bushyhead responded by giving Agent John Q. Tufts at the Union Agency in Muskogee a list of the names of recalcitrant stockmen. Tufts thereupon began a sequence of letters that requested authority of Commissioner of Indian Affairs Hiram Price to dispatch troops from Fort Gibson to remove the ranchers.[14]

13. D. W. Lipe to Hon. D. W. Bushyhead, April 4, 1881, File: Cherokee—Strip (Tahlequah), 1880–1881, IAD, OHS. Morris and McReynolds, *Historical Atlas*, Map 41.

14. J. G. Schrimsher to Dear Dennis, June 21, 1881, File: Cherokee—Strip (Tahlequah), 1880–1881, IAD, OHS. D. W. Bushyhead to J. Q. Tufts, July 20, 1881, ibid. H. Price to The Secretary of the Interior, August 22, 1881 (copy), ibid.

No. 82

GRAZIER'S LICENSE.

CHEROKEE NATION,

OFFICE NATIONAL TREASURER,

Caldwell Kans Oct 14 1881

By virtue of authority given me by an act of the National Council, approved December 1st, 1880, to supervise the collection of the Revenue of the Nation derivable from that portion of the country not included within the Nation as organized by law, but remaining in the possession of the Nation for the profits accruing from such possession, until sold and occupied in accordance with the provisions of the 16th Article of the Cherokee Treaty of 1866, I hereby acknowledge to have this day received, for the Treasury of the Cherokee Nation, the sum of Sixteen Hundred *Dollars from* Gregory Eldridge & Co *citizens of the United States, the said sum being paid by said party, and received by me, as payment made to said Nation, for grazing* Cattle *upon the lands designated, to the number of* four thousand *head, and no more, for the period of time included within* first *day of* Sept *1881, and the* first *day of* Sept *1882.*

Authorized Agent for Treasurer Cherokee Nation.

Treasurer of Cherokee Nation.

Grazier's license issued by the Cherokee Nation, 1881

On June 1, Schrimsher hired Joseph G. McCoy, the Illinois livestock dealer who had opened the cattle market at Abilene, Kansas, in 1867, to assist in tax collection. Wise in the ways of cattlemen, McCoy quickly assessed conditions on the Outlet and sent his views to Bushyhead. He had talked with many of the men who grazed stock on Cherokee land and had reached the conclusion that they would never pay for the privilege unless forced to do so. Word that the Principal Chief had provided the United States Indian Agent with the names of delinquents had caused "a ripple of excitement among graziers," but few felt that the news warranted payment of taxes. Since a show of federal force had not been made immediately, cattlemen had grown "jubilant, defiant, and insulting," branding the Cherokees' treasury agents "frauds, liars, and confidence men" and treating them "as mendicants soliciting charitable contributions." Those who had paid the tax regretted their promptness, in the belief that they had been "swindled out of their money upon false representation." McCoy advised Bushyhead to request help for his agents. "If their only support in the discharge of thier [*sic*] unpleasant duty," he said,

> is to be glittering promises on paper from the Int. Dept. which are ignored and broken without a pretense of fidelity to agreements; then independent dignity would suggest that all attempts to collect grazing taxes be abandoned and your Agents recalled.[15]

15. J. G. McCoy to Hon. D. W. Busheyhead [*sic*], August 22, 1881, File: Cherokee—Strip (Tahlequah), 1880–1881, IAD, OHS.

McCoy's report undoubtedly troubled Bushyhead, and the news he received from William A. Phillips, counsel for the Cherokee delegation then in Washington, to the effect that unassigned federal troops were scarce as a result of the Apache wars in Arizona, did not ease his mind. The Principal Chief, sensing the futility of the situation, could only telegraph Secretary of the Interior Samuel J. Kirkwood to inform him of the presence of nontaxpaying cattlemen on Cherokee land.[16]

Officials in Washington were not insensitive to the plight of the Cherokees. While Lipe and his deputies attempted to collect the grazing tax and Bushyhead sought to coordinate his nation's efforts from Tahlequah, Commissioner Price worked to unravel the complexities of the problem and to define the federal government's role in solving it. After careful consideration of the facts in the case, Price presented his views to Secretary Kirkwood. Acknowledging that the grazing tax had been sanctioned by the Department of the Interior, Price reasoned that delinquent cattlemen were intruders on Cherokee land and, therefore, liable to removal under the provisions of the Treaty of 1866. But Agent Tufts, the man upon whom fell the responsibility for evicting interlopers, controlled a police force that was "manifestly inadequate for the undertaking." Price therefore recommended the use of federal troops. Kirkwood, who had received Bushyhead's telegram just two days earlier, sent orders to the War Department on August 26 to remove "delinquent cattle graziers" from the Cherokee Outlet. He then notified the Principal Chief of his action. Price wired Tufts on August 27 that troops would be made available to assist in the eviction of trespassers.[17]

Kirkwood's action came too late to affect revenue collection in 1881. Before the troops arrived in the area the season had ended and herds were turned loose on the Outlet for the winter. Tufts and fed-

16. W. A. Phillips to D. W. Bushyhead, August 21, 1881, File: Cherokee—Strip (Tahlequah), 1880–1881, IAD, OHS. Bushyhead to Kirkwood, August 24, 1881 (telegram), Letters Received, 1881–1907, Record Group 75, Records of the Bureau of Indian Affairs, National Archives, Washington, D.C. Hereafter cited as Letters Received, RG 75, RBIA.

17. Kappler, *Indian Affairs,* II, 948. H. Price to The Secretary of the Interior, op. cit. S. J. Kirkwood to Commissioner of Indian Affairs, August 26, 1881, Letters Received, RG 75, RBIA. Kirkwood to Bushyhead, August 26, 1881 (telegram; copy), ibid. H. Price to Tufts, August 27, 1881 (telegram; copy), File: Cherokee—Strip (Tahlequah), 1880–1881, IAD, OHS.

eral troops could not sort the cattle of nontaxpayers from among the thousands of head of livestock grazing the pasture beyond the Arkansas, so the efforts of Lipe and his deputies were frustrated for that year. Yet, despite the obstinacy of some ranchers, the Cherokee Nation managed to collect over $21,000 in grazing taxes during 1881—nearly three times the amount gathered in 1880. More significant than the trebled income was the fact that the federal government had demonstrated to cattlemen its willingness to act in behalf of the Cherokees under the 1866 treaty. Thus, in January, 1882, ***The Cherokee Advocate,*** official voice of the Tahlequah government, could state with some assurance that

> those persons who have cattle grazing on our Strip might as well pay their taxes, and save trouble. Uncle Sam stands by the Cherokees in this matter, and those stockmen who have stock on the Cherokee Strip, and who are kicking against paying taxes to the Cherokee authorities, are simply cutting their own throats—in other words, "no pay, no stay." [18]

Subsequently, the Indians lowered their assessments to forty cents per head annually for cattle over two years old and twenty-five cents per head for all others, but they did not lessen their efforts to collect these reduced taxes. As a result, during the 1882 season Treasurer Lipe issued tax receipts totaling more than $41,000.[19]

From the perspective of the Cherokees, the grazing tax was at last a success—a welcome one, since it provided revenues the National Council appropriated for tribal educational institutions. Still, the Nation had not altered its intention to sell the Outlet, and the federal government's readiness to assist the tribe in 1881 must have been

18. J. G. McCoy to Hon. D. W. Busheyhead [*sic*], September 1, 1881, File: Cherokee—Strip (Tahlequah), 1880–1881, IAD, OHS. *The Cherokee Advocate* (Tahlequah), February 6, 1885; January 6, 1882.

19. Testimony of Andrew Drumm before the Senate Committee on Indian Affairs, January 8, 1885, U.S., Congress, Senate, 49th Cong., 1st sess., 1885, Report 1278, VIII, Part I, p. 77. *The Cherokee Advocate* (Tahlequah), February 6, 1885.

viewed by many as a sign that federal officials would indeed complete the transfer arranged for in the Treaty of 1866. By 1882, however, cattlemen had a vested interest in the Outlet, and some were actively seeking ways in which to secure grazing rights and at the same time circumvent the tax. A few spoke of leasing the land west of the Arkansas, but such steps required cooperation and organization, and as yet stockmen lacked these requisites for effective action. The troubles of the 1882 season, however, provided ranchers with the opportunity to coordinate their activities.[20]

20. D. H. Ross, R. M. Wolfe, and W. A. Phillips to S. J. Kirkwood, January 24, 1882, Letters Received, RG 75, RBIA. See House Report 3768, op. cit., pp. 22–25. J. G. McCoy to Hon. D. W. Busheyhead [*sic*], September 1, 1881, op. cit.

Herd on the Cimarron River in the 1880s

II. THE ASSOCIATION

By comparison to their counterparts in Texas and on the Northern Plains, cattlemen on the Cherokee Outlet faced an anomalous situation. They were little more than tenants on Cherokee land, enjoying grazing privileges only so long as they paid taxes to the tribe. Reluctance to pay might result in removal by federal troops, yet compliance with Cherokee law rarely brought security on the range. Cattlemen constantly faced the threat of encroachment by outsiders. Homeseekers in ever-increasing numbers were gathering along the Kansas line in anticipation of government action to open Indian Territory to settlement, but initially, the Outlet ranchers' greatest difficulties were caused by Kansas stockmen who paid no taxes to the Cherokees. These men grazed herds near the boundary, moving south to take advantage of Outlet pasture and returning north when Indian tax agents appeared. Thus, land near the line was subject to periodic overgrazing that diminished its value to the responsible ranchers who had paid for the right to use it.[1]

In addition, timber thieves from Kansas were a constant menace to the land, for they cut and removed wagonloads of the Outlet's valuable cedar. Their intrusions angered the Cherokees, and cattlemen feared that the depredations, if unchecked, might prompt tribal officials at Tahlequah to close the Outlet to occupancy in any form.

1. A. Drumm to Hon. D. W. Lipe, February 14, 1882, CNP, WHC. C. M. McClellan to Hon. D. W. Lipe, March 10, 1882 (copy), File: Cherokee—Strip (Tahlequah), 1882, IAD, OHS.

Thus, the problems posed by the theft of timber became the ranchers' as well as the Cherokees'.[2]

Anomaly was also inherent in the peculiar grazing methods the Outlet cattlemen developed. Because of the unusually large number of herds west of the Arkansas, ranchers were forced to employ many cowboys throughout the year. While hired hands were added seasonally on the open range for round-ups, branding, and the drive to market, Outlet stock raisers signed on men primarily as line riders. Their principal duties were to keep the herds separated and confined to specific pastures. Wages for line riders entailed a substantial increase in operating costs for cattlemen already burdened with grazing taxes. Moreover, line riding established wide dividing grounds between herds. By maintaining these separate pastures, ranchers not only increased an expense that was unnecessary but also used the land inefficiently.[3]

To obviate these costs, Outlet cattlemen turned to fencing with barbed wire to separate the pastures. This inexpensive fencing was developed in 1873 by Joseph Farwell Glidden, an Illinois farmer who elaborated on the ideas of others to make production of the wire commercially practical. Before the decade was out, more than 86,900 tons of barbed wire had been sold by various manufacturers, most of it for use on the Great Plains, where traditional fencing materials like timber and stone were in short supply. Again, anomaly was an element of life on the Cherokee Outlet, because elsewhere farmers, not cattlemen, were the prime purchasers of barbed wire. Western stockmen customarily opposed fencing—especially wire fencing—because it was a hazard to cattle and menaced the continuance of open-range ranch-

2. J. G. McCoy to Hon. D. W. Busheyhead [*sic*], September 1, 1881, File: Cherokee—Strip (Tahlequah), 1880–1881, IAD, OHS.

3. J. W. Strong to Hiram Price, October 10, 1881, U. S., Congress, Senate, 48th Cong., 1st sess., 1883, Executive Document 54, IV, p. 128; B. H. Campbell to Hon. H. M. Teller, January 2, 1883, Letters Received, Special File 9, RG 75, RBIA.

ing, yet on the land beyond the Arkansas it was heralded as a remedy for a variety of economic ills. It has been stated that in some areas the barbed-wire fences were used to keep livestock out of pastures rather than in them, but this view does not apply to the problems of the Cherokee Outlet.[4]

By fencing their grassland, Outlet ranchers hoped to reduce the number of men on their payrolls, utilize the maximum amount of available pasture, and prevent intrusion by nontaxpaying cattlemen. Construction of barbed-wire fences would also assist Cherokee tax collectors, who would no longer have to scour the countryside searching for wandering herds of cattle. And if fencing impeded the activities of timber thieves or made their arrest easier, then so much the better for the stockmen, who counted on the Cherokees' friendship in permitting them to remain on the land. They applied to the representative of the Cherokee Nation with whom they had closest contact—D. W. Lipe—for permission to erect enclosures.[5]

One of the first cattlemen to approach the Cherokee Treasurer on the matter of fencing was Andrew Drumm, a prominent rancher who had driven longhorns from southern Texas to Kansas in 1870 and was reputedly the first to have grazed stock on the Outlet. Drumm maintained that an increase in the Cherokee tax rate would constitute unjust discrimination. He outlined, early in 1882, the conditions he was willing to meet:

> In fencing . . . I propose to be governed by the Cherokee Authorities; that whenever, if ever, said "Strip" is transferred to the U.S. Government, or in any other manner disposed of by your Nation, I have the privilege of moving or doing

4. Walter Prescott Webb, *The Great Plains*, pp 298–99, 309. For the traditional view of the effect of barbed wire on the cattle industry, see Webb, p. 317, and Henry D. and Frances T. McCallum, *The Wire that Fenced the West.*

5. B. H. Campbell to Hon. H. M. Teller, January 2, 1883, op. cit.

Andrew Drumm, director of the Cherokee Strip Live Stock Association

> whatever else I see fit to do with my fencing; that for the privilege of so being allowed to graze my stock on the Strip, I propose to pay the Nation Two Thousand Seven Hundred Dollars per year, or just such amount of tax as is now collected upon each head. Providing I am not unjustly discriminated against after I have fenced the range upon which my cattle or stock run; Provided, further, That when I am removed from said Strip I will leave all the posts, and other improvements excepting the wire, which are to go to the Cherokee Nation for its use and benefit.[6]

Lipe forwarded Drumm's application to Principal Chief Bushyhead with his own endorsement of the plan. Lipe was unaware of "any law with which the privilege [of fencing], if granted, would conflict." Enclosures, he believed, "would ultimately be the means . . . of collecting a much larger revenue than is now collected—and in the end compel these persons who refuse to pay tax to either pay or remove from off the lands." [7]

Formal permission for cattlemen to fence Outlet pasture was not immediately forthcoming from tribal officials at Tahlequah, but the delay had little effect on the ranchers' wire-stringing activities. Lipe issued no official enclosure permits, but when his office received applications for them, he replied simply that he had "no objections" to fencing. Many ranchers took him at his word; others, lacking the temerity to risk the Cherokees' wrath should Cherokee officials subsequently announce an antienclosure policy, hired Cherokee citizens to build fences for them. The question of who actually constructed the fences had no bearing on their legality, however, since in every case they were unauthorized. Clearly, Cherokees were in tacit agree-

6. L. S. Records, "The Recollections of a Cowboy in the Seventies and Eighties: The Personal Observations of L. S. Records," arranged and prepared by Ralph H. Records, 481, manuscript, WHC. A. Drumm to Hon. D. W. Lipe, February 14, 1882, op. cit.

7. D. W. Lipe to Hon. D. W. Bushyhead, February 14, 1882, CNP, WHC.

ment on the value of enclosures, viewing them as a means of facilitating tax collection. In Lipe's opinion, ranchers strung barbed wire "at their own risk and will be responsible for the taxes on all cattle held therein." [8]

In May, 1882, the Interior Department became interested in the fencing on the Outlet. W. W. Woods, a rancher from Kiowa, Kansas, complained to Commissioner of Indian Affairs Hiram Price that he had been evicted from the Outlet when Andrew Drumm enclosed the land upon which Woods had been paying taxes to the Cherokees for three years. Drumm and his partner, Woods claimed, had "six mule teams in the beautiful cedar timber south of the Cimmeron [*sic*] River cutting and hauling posts for their fence." He warned that "in twelve months from now every stick of timber in this Indian Territory that will make a fence post will be cut down." [9]

Officials of the Bureau of Indian Affairs were unconcerned about the range dispute between Woods and Drumm. What arrested their attention was Woods's disclosure of the extent of cattlemen's timber-cutting operations on Outlet land. Acting Commissioner E. S. Stevens reminded the Cherokee delegates, Daniel Ross and R. M. Wolfe, that Tahlequah had "repeatedly complained to this office on this very subject" and suggested that the problem be brought to the Nation's "immediate attention in order that steps may be taken to arrest the movement." Ross and Wolfe apprised Bushyhead of the situation, warning him that they anticipated intervention "in some shape by the United States government." The Principal Chief turned to Lipe for information upon which to base his response.[10]

In his response, Lipe reviewed his policy toward fencing in the

8. D. W. Lipe to C. M. McClellan, March 13, 1882 (copy), File: Cherokee—Strip (Tahlequah), 1882, IAD, OHS. D. W. Lipe to Hon. D. W. Bushyhead, May 29, 1882 (copy), ibid.

9. W. W. Woods to H. Price, May 2, 1882, Letters Received, RG 75, RBIA.

10. E. S. Stevens to D. H. Ross and R. M. Wolfe, May 25, 1882, File: Cherokee—Strip (Tahlequah), 1882, IAD, OHS. Danl. H. Ross and R. M. Wolfe to Hon. D. W. Bushyhead, May 25, 1882, ibid.

Outlet, asserting that "objections (if any) will come from parties along the line of Kansas, who will be debarred from using our country as a free grasing [*sic*] ground for their stock." Woods, according to Lipe, was one of those who ran cattle "on both sides beating his own state and the Cherokee Nation too." The amount of timber cut by cattlemen was negligible when compared to the vast quantities of cedar stolen each year by Kansans. Lipe believed that the barbed-wire fencing would "protect the timber to as great and the same extent that it will protect the range." [11]

Bushyhead forwarded Lipe's views to Ross and Wolfe in Washington. Cattlemen learned, probably from Lipe, of the federal government's interest in the question, and in June several prominent Outlet ranchers wrote to Secretary of the Interior Henry M. Teller to allay bureaucratic fears about the status of Cherokee land. They praised the installation of barbed-wire fencing as a positive good for stockmen and Indians alike. Others attested to the injustices done Outlet cattlemen who paid the tax, by nontaxpaying Kansans, and pointed to fencing as the most direct solution to their problems. Officials in Washington probably assigned greater validity to the information provided by the Cherokee delegation than to the arguments of ranchers, but in any case, the complaint of W. W. Woods of Kiowa, Kansas, was laid to rest. Unauthorized fencing continued in the Outlet, although a bill was presented in the Cherokee National Council to prohibit Cherokee citizens from building fences for cattlemen. Bushyhead vetoed the measure, and there matters stood until the winter of 1882.[12]

11. D. W. Lipe to Hon. D. W. Bushyhead, May 29, 1882, op. cit. D. W. Lipe to Hon. D. W. Bushyhead, June 5, 1882, File: Cherokee—Strip (Tahlequah), 1882, IAD, OHS. D. W. Lipe to Hon. D. W. Bushyhead, May 29, 1882, op. cit.

12. W. P. Boudinot to D. H. Ross and R. M. Wolfe, June 16, 1882, File: Cherokee—Strip (Tahlequah), 1882, IAD, OHS. Jess Evans, A. G. Evans, et al. to Hon. Mr. Teller, June 14, 1882, Letters Received, RG 75, RBIA. Deposition of W. H. Harrelson, June 28, 1882, ibid. H. Price to The Secretary of the Interior, December 28, 1882 (copy), CNP, WHC.

Christmas Day of 1882 passed quietly in the Cherokee Outlet. Cattlemen who spent the holidays with their families in Wellington, Kiowa, Caldwell, or any of a dozen other settlements near the Kansas border had no warning that in Washington a storm was brewing that would thrust them headlong into a decade of sparring with the federal government to defend their occupancy of grazing land in Indian Territory.

The government's first challenge to the cowman's last frontier was formulated on December 28, 1882, when Commissioner Price wrote to Secretary of the Interior Henry M. Teller urging "an end . . . to . . . unauthorized settlement and improvement" of the Outlet. Ironically, Price was motivated by information received from C. M. Scott, a taxpaying rancher from Arkansas City, Kansas. Scott claimed that part of his Outlet range had been illegally fenced by a Cherokee citizen acting in behalf of the Pennsylvania Oil Company. The company's wire, he said, threatened to close United States mail routes across Indian Territory. Price cited an interpretation of the 1866 treaty, issued two years earlier by the Attorney General, to support his contention that fencing on the Outlet was in violation of federal law. He enjoined Teller to use military force to remove enclosures and thereby protect "the rights reserved to the Government." [13]

13. H. Price to The Secretary of the Interior, op. cit.

Teller did not hesitate to act. On December 29 he sent Price's letter, together with his own instructions, to Secretary of War Robert Lincoln. Upon written request from Agent John Tufts, Lincoln was

Typical cowhand accommodations in the Outlet during the 1880s

to dispatch troops to destroy "all improvements of every character" on the Cherokee Outlet, unless they were first removed voluntarily by the cattlemen who had constructed them. Notified of Teller's decision, Price ordered Tufts to inform ranchers that fences, corrals, and dwellings should be dismantled within twenty days, or federal troops would be sent to carry out the order.[14]

Shocked ranchers replied quickly to Teller. B. H. Campbell of Wichita, who later managed the XIT spread, defended cattlemen's right to enclose pasture in the Outlet as a practice beneficial to all concerned. No stockman, he said, believed fences gave him permanent claim to Cherokee land, because each was prepared to leave the Outlet whenever the federal government settled Indians on it or opened it to homesteaders. Campbell urged Teller to make a thorough investigation of the situation before executing the fence-removal order. Other cattlemen, insisting that they built fences solely "for the more economical use of the land" and not to create "permanent ranches or . . . permanent improvements," asked flatly that the order be revoked.[15]

Secretary Teller was not the only recipient of such letters. Because his request was responsible for the Army's orders to destroy fences, Agent John Tufts was another target for the correspondence of Outlet ranchers. The cattlemen cited the $100,000 they had invested in fencing, and they demanded a fair hearing before their investment should be destroyed. They had acted with the assent—if not the consent—of the Cherokees and were unaware of having violated any law. Tufts relayed their views to Price, with the recommendation that a formal hearing be conducted in Washington to satisfy the concerned

14. H. M. Teller to The Secretary of War, December 29, 1882 (copy), CNP, WHC. H. Price to John Q. Tufts, December 30, 1882 (copy), ibid.

15. B. H. Campbell to Hon. H. M. Teller, January 2, 1883, op. cit. E. M. Ford, A. Drumm, et al. to The Secretary of the Interior and Commissioner of Indian Affairs, January 2, 1883, Senate Executive Document 54, op. cit., p. 135.

parties. He told the Commissioner that he planned to investigate conditions in the Outlet personally and to report his observations to Washington. Then, on January 4, Tufts complied with Price's earlier instructions and ordered cattlemen to remove all improvements from the Outlet by the first of February.[16]

The ranchers responded with fresh arguments. They branded Tufts's action unreasonable and peremptory. Without basic improvements, they argued, the range cattle industry could not survive in the Outlet. The threatened federal action would effectively end ranching there, but because of the 1866 treaty, the land was "unavailable for every other purpose than grazing." Judging that the government had rendered the Outlet useless to red men as well as white, the cattlemen urged suspension of the directive.[17]

In answer to a question raised by Secretary of War Lincoln, Price outlined the legal basis for the fencing order. While the use of federal troops to remove improvements from the Outlet would admittedly be unlawful, the Commissioner justified it as a measure necessary to protect Cherokee lands from intruders, an obligation placed on the United States by Article XXVI of the 1866 treaty. Construction of unauthorized fencing, Price believed, constituted intrusion. He found precedent for this view in an 1834 statute that, according to his interpretation, would make each offending cattleman subject to a fine of one thousand dollars.[18]

Price's analysis of the problem led him to conclude that, while the Cherokees officially welcomed cattlemen in the Outlet and profited from their presence, the tribe did not condone fencing because it created "a system of apparent ownership." Enclosures, he reasoned,

16. John Q. Tufts to H. Price, January 3, 1883, Senate Executive Document 54, op. cit., p. 138. John Q. Tufts to H. Price, January 4, 1883, ibid.

17. F. L. Underwood to Hiram Price, January 5, 1883, Senate Executive Document 54, op. cit., p. 139. See also John L. McAtee to Hon. Arthur P. Gorman, January 5, 1882, ibid., p. 138.

18. Robert T. Lincoln to The Secretary of the Interior, December 30, 1882, Senate Executive Document 54, op. cit., p. 131. H. Price to Secretary of the Interior, January 6, 1883 (copy), CNP, WHC.

concentrated vast holdings "in the hands of a few moneyed individuals and corporations to the exclusion of many of the less favored." Thus, in the Commissioner's mind, the situation smacked not only of illegality but also of immorality.[19]

On January 9 Tufts wired Price to ask if the removal order had been rescinded. The Commissioner replied that it had not and that matters were currently in the hands of the military. When ranchers learned that the government had not altered its position, their anxiety grew. Some cattlemen had constructed drift fences to prevent winter losses, and they feared that their herds might be endangered if the February 1 deadline were to be met. Congressmen received veritable avalanches of mail from constituents with investments in Outlet ranching. Feverishly, the cattle interests began work toward creating an effective lobby. Only a few small ranchers who complained that fencing by large cattle companies crowded them from the range supported Teller and Price.[20]

In mid-January, Price postponed the deadline and directed Tufts to conduct an investigation of conditions in the Outlet. Specifically, Tufts was to secure information about the quantity of fencing on Cherokee land and obtain the names of those who owned it. In his report, he was to estimate the amount of timber cut from Indian land for fence posts and to evaluate the effect of enclosures on "legitimate trade and travel" and mail routes.[21]

Tufts filed his report on March 1. Nineteen individuals or companies, he found, had constructed 959 miles of barbed-wire fence on Outlet land. Fences were well supplied with gates and constituted no barrier to travelers or federal mail wagons. Timber had indeed been

19. H. Price to Secretary of the Interior, op. cit.

20. Tufts to Commissioner of Indian Affairs, January 9, 1883, and H. Price to Tufts, January 10, 1883 (telegrams), Senate Executive Document 54, op. cit., pp. 140–41. F. M. Cockrell to Henry M. Teller, January 10, 1883, and Nicholas Ford to The Secretary of the Interior, January 12, 1883, ibid., p. 141. J. A. McPhee to Henry M. Teller, January 12, 1883, ibid., pp. 141–42.

21. H. Price to John Q. Tufts, January 16, 1883, Senate Executive Document 54, op. cit., p. 142; Questions Issued by Secry of Interior [*sic*] to Agent Tufts to investigate (undated note in handwriting of Charles H. Eldred), File: Cherokee Strip Live Stock Association (Section X), IAD, OHS.

cut in large quantities, but Tufts attributed the destruction to thieves from Kansas. Cattlemen, he wrote, "neither cut timber themselves, nor do they permit anyone else to do so." [22]

Tufts recommended that ranchers be allowed to continue fencing only if they first obtained permission from the Cherokees, but, he wrote, they should understand that the wire could be removed at the discretion of the Department of the Interior. He reported that most ranchers in the Outlet were well-intentioned men who relied on fences to protect their range from nontaxpayers' cattle grazing along the Kansas line. These interlopers were vehement in their opposition to fencing because it impeded their evasion of Cherokee treasury agents. If honest cattlemen continued to enclose the range, Tufts concluded, "the Cherokee Nation will collect double the tax." [23]

Undoubtedly, Bushyhead and Lipe agreed with the substance of Tufts's report, but not all Cherokees shared their views. R. M. Wolfe and Robert B. Ross, tribal delegates to Washington, and William A. Phillips, Cherokee attorney, doubted the accuracy of Tufts's findings. In a letter to Teller, these dissidents argued that it was impossible for cattlemen to string 959 miles of wire without using Outlet timber for posts. They estimated that approximately two hundred thousand posts were needed to set that extent of fence, and they charged that the ranchers had cut them from the region's valuable cedar trees without permission from, or compensation to, the Cherokee Nation. The delegates urged seizure by the federal government of all fencing as indemnification for the Cherokees' losses and concluded with a warning to Teller of "the danger of rescinding your order." [24]

Before passing judgment on the issue, Teller reviewed Tufts's

22. John Q. Tufts to Hon. H. Price, March 1, 1883, Senate Executive Document 54, op. cit., p. 149.

23. John Q. Tufts to Hon. H. Price, op. cit.

24. R. M. Wolfe, Robert B. Ross, and William A. Phillips to H. M. Teller, March 12, 1883 (copy), CNP, WHC.

Cowhands and chuckwagon

report and correspondence from ranchers who favored a permissive enclosure policy. On March 16 he announced his decision in a letter to Price:

> No further fences will be permitted to be constructed on these lands. Those constructed will not be permitted to remain except with the consent of, and under proper and satisfactory arrangements with, the Cherokee national authorities, to be secured within reasonable time to be fixed by you; failing in which, the order heretofore given for the removal of the fences will be at once enforced.[25]

By the time Teller issued his pronouncement, Outlet cattlemen had moved decisively to protect their interests. Three years earlier, ranchers had founded a loosely knit "convention" to expedite round-ups and facilitate a policy of mutual assistance. In January, 1883, responding to Washington's initial fence-removal order, members of the convention met in Topeka, Kansas, to strengthen their organization. The resulting Cherokee Strip Stockmen's Association at first confined its activities to drafting letters of protest to the Department of the Interior. When it became clear that Teller and Price would not be easily moved from their position on the fencing question, the ranchers prepared to take stronger action.[26]

Meeting in Caldwell on March 6, the cattlemen appointed a committee of nine to prepare a constitution, bylaws, and a charter of incorporation for a new organization. Within forty-eight hours, they completed and signed the charter, thus bringing to birth the Cherokee Strip Live Stock Association. According to its charter, the association was incorporated under Kansas law for a term of forty years to promote "improvement of the breed of domestic animals by the im-

25. H. Price to The Secretary of the Interior, March 14, 1883, Senate Executive Document 54, op. cit., pp. 151–52. H. M. Teller to The Commissioner of Indian Affairs, March 16, 1883, ibid., p. 152.

26. Testimony of Benjamin S. Miller before the Senate Committee on Indian Affairs, January 9, 1885, U. S., Congress, Senate, 49th Cong., 1st sess., 1885, Senate Report 1278, Part I, p. 80. There is some confusion over the date of the cattlemen's first meeting. Miller recalled that the Cherokees first collected taxes in the Outlet in 1880 after the initial meeting. He remembered the tax collector's name as Bell, but, according to *The Cherokee Advocate* (Tahlequah), February 6, 1885, Bell served as Special Tax Collector only in October, 1879. This indicates that the first meeting may have occurred in 1879, but Edward Everett Dale, *The Range Cattle Industry*, p. 136, accepts the 1880 date. J. A. McPhee to Henry M. Teller, January 12, 1883, op. cit.

portation, grazing, breeding, selling, bartering, and exchange thereof" in places "most advantageously located." The nine committeemen who drafted the document became directors of the association, and they elected as its president Benjamin S. Miller of Caldwell, one of their number and chairman of the old convention.[27]

27. Charter of the Cherokee Strip Live Stock Association, U. S., Congress, Senate, 48th Cong., 2d sess., 1884, Executive Document 17, I, p. 149.

The association reserved in its bylaws the right to "purchase or lease any and all parcels or tracts of land, wheresoever situated, as may be necessary" to conduct its business. Ranchers who occupied undisputed range in the Outlet were eligible for membership, which required payment of ten dollars. Cattlemen outside the Outlet might become honorary members upon the recommendation of the directors. Individuals and corporations alike were to have a single vote in the association's balloting, and the bylaws assured equal consideration for large and small outfits. The association's secretary was empowered to keep records of all range transfers and lists of members' brands. Disputes arising between members were to be settled by a board of arbitration consisting of three ranchers appointed by the directors. Decisions of this arbitration body could be appealed to the directors, whose verdict was to be final. In short, the machinery of the association was to function as simply and as efficiently as possible.[28]

28. Bylaws of the Association, Senate Executive Document 17, op. cit.

Like the cattlemen who were its members, the Cherokee Strip Live Stock Association was, from the very first, mired in a strange, ambiguous situation. The phenomenon of the stockmen's organization flourished throughout Western cattle country in the 1870s and 1880s, but, as had been noted, almost without exception *these groups were formed to regularize range practices—branding, round-up, destruction of predators, disposition of mavericks, prevention of rustling, and the like—that had once been conducted by informal agreements known collectively as* **cow custom.** *To these the Cherokee Strip Live Stock Association bore no resemblance. It was not born spontaneously in bucolic environs as the corporate offspring of like-minded men working amiably at a leisurely pace. Rather, it was thrust into existence as a desperate measure taken by troubled men acting out of an urgent sense of the need to defend themselves against the encroachments of a bureaucracy that they viewed as bent on regulation. And that regulatory aspect of the government, too, was atypical in the period.*[29]

In the years following the Civil War the federal government adopted a flexible laissez-faire attitude toward entrepreneurs who were developing a mushrooming industrial economy in the East. Yet, cattlemen on the lands beyond the Arkansas enjoyed no such freedom. Possibly the difference in the government's attitude toward them lay in the Westerners' relationship with the Indians, themselves the victims of unyielding federal paternalism. In any case, the business operations of ranchers in the Outlet were subject to an intense scrutiny not extended to businessmen in the East. It is perhaps too much to say that they organized to guard the principles of free enterprise, but certainly their efforts produced that effect. Cattlemen in the Cherokee Outlet did only what Washington forced them to do. The informal convention of an earlier day had functioned well enough to fit the needs of men in an economy that depended only on grass and market. Faced with a real threat to their economic security, ranchers

29. See Arrell M. Gibson, "Ranching on the Southern Great Plains," *Journal of the West*, 6:1 (January 1967), 147–48.

sought strength in numbers and safety in bylaws and charters. Had they been allowed to function as in the past, an association would have been unnecessary. They incorporated to save themselves, but in the end, they succeeded only in postponing the inevitable.[30]

30. William W. Savage, Jr., "Barbed Wire and Bureaucracy: The Formation of the Cherokee Strip Live Stock Association," *Journal of the West*, 8:3 (July 1968), 406, 412.

III. THE LEASE

Chief Dennis Bushyhead was in Washington on March 16, 1883, when Secretary Teller announced the decision that forbade enclosures in the Outlet. On March 20 he met with Hiram Price and assured the Commissioner that he would call an extra session of the Cherokee National Council as soon as possible to discuss a new arrangement with Outlet cattlemen. Bushyhead and Price agreed that Agent Tufts should furnish police to enforce the enclosure ban until the council acted. After the meeting, Bushyhead sent word to John F. Lyons, a Fort Gibson attorney employed by the Cherokee Strip Live Stock Association, that the federal government would not act immediately to remove ranchers' fences. Lyons relayed the information to a director of the association, Charles H. Eldred of Medicine Lodge, Kansas, and prepared to attend the extra council session, which would convene in Tahlequah when Bushyhead returned from Washington.[1]

Directors Eldred and Andrew Drumm represented the association at the council meeting. With Lyons and E. C. Wilson, another attorney, they went to Tahlequah, Eldred recalled later, to consult "with members of the council regarding what they thought would be advantageous to the nation and to other parties concerned." The association's bylaws, of course, mentioned the possibility of leasing

1. H. Price to John Q. Tufts, Esq., March 21, 1883, U. S., Congress, Senate, 48th Cong., 1st sess., 1883, Executive Document 54, IV, p. 153. John F. Lyons to Charles H. Eldred, March 24, 1883, File: Cherokee Strip Live Stock Association (Section X), IAD, OHS.

Charles H. Eldred, director of the
Cherokee Strip Live Stock Association

land, as in all probability this matter was uppermost in the minds of those who traveled to the Cherokee capital. It has been argued that the idea of leasing was the primary factor that influenced the formation of the association. The notion occurred to cattlemen as more than an afterthought, to be sure, but it was certainly not the association's ***raison d'être.*** *If previous experience held any influence, and if incorporation depended upon the certainty of obtaining a lease, cattlemen would have developed nothing beyond the informal organization of the old convention. Early efforts to lease the Outlet had failed, and in the spring of 1883 there was no reason to believe that success was imminent. Undoubtedly there were factors other than leasing to be considered by the association.*[2]

The idea of leasing the Cherokee Outlet was not new. Perhaps the first attempt to acquire the land beyond the Arkansas on a contractual basis was made in 1875 by two Kansans named Walker and Hughes. They applied to the Department of the Interior for a three-year lease and offered to pay $1,000 per year. The department referred them to the Cherokee delegation in Washington, and a year later a bill authorizing a seven-year lease in their names reached the tribal council. Those who recognized the value of the Outlet forced the bill's withdrawal. Then, during the summer of 1880, shortly after the creation of the convention, two Texas cattlemen, Patrick Henry and D. J. Miller, representing a group of Waco ranchers, offered the Cherokees $185,000 for rights to Outlet range. The men boasted of having over $1 million in capital resources, and they brought $85,000 to Tahlequah as evidence of good faith. They were questioned by tribal representatives but became evasive when asked how they planned to use

2. Testimony of Charles H. Eldred before the Senate Committee on Indian Affairs, January 13, 1885, U. S., Congress, Senate, 49th Cong., 1st sess., 1885, Senate Report 1278, p. 151. Edward Everett Dale, *The Range Cattle Industry*, p. 139.

the land. Members of the convention entered bids against the Texans, but the Cherokees rejected all offers. A year later, Joseph G. McCoy told Bushyhead that if Cherokees "would authorize the leacing [**sic**] in tracts of its outlet . . . on ten year terms . . . at prices graduated according to quality of grazing location and water . . . it would be more satisfactory to graziers and double the annual revenue" of the tribe.[3]

Despite the cattlemen's enthusiasm, the lease idea was not well received in other quarters. Senator Preston B. Plumb of Kansas denounced it as "bad policy on general grounds" and predicted that a "monopoly of that kind would break down of its own weight." He was understandably concerned about the reaction of would-be settlers among his constituents who wished to see the Outlet added to the public domain. The position of the Cherokees was less easily determined. Officially, the government at Tahlequah declined all offers to lease their land. The tribal delegation in Washington informed the Secretary of the Interior in January, 1882, of cattlemen's efforts to obtain leases, commenting that they "doubted the expediency of creating a monopoly of the grazing privileges, which might lead to complaints against our administration of the matter." While the Cherokees clearly wanted to derive income from the Outlet, they were particularly sensitive to the possibility of arousing conflict with the federal government. They continued to consider selling the land beyond the Arkansas, but they first had to confront a problem of a more immediate nature. The federal government, they believed, still owed the tribe for lands ceded in 1866 for the settlement of friendly Indians.[4]

3. Testimony of Dennis W. Bushyhead before the Senate Committee on Indian Affairs, May 21, 1885, Senate Report 1278, op. cit., II, p. 60. Joint Committee to take in Consideration the Chief Vita Message to bill No. 5, in relation to the grazing privilege west of 96°, July 14, 1880, File: Cherokee—Strip (Tahlequah), 1880–1881, IAD, OHS. J. G. McCoy to Hon. D. W. Busheyhead [*sic*], September 1, 1881, ibid.

4. J. B. Plumb to C. Schurz, January 4, 1881, Letters Received, RG 75, RBIA. D. H. Ross, R. M. Wolfe, and W. A. Phillips to S. J. Kirkwood, January 24, 1882, ibid.

Tahlequah in 1888

During the summer of 1879 the Office of Indian Affairs conducted an appraisal of the Outlet, estimating the worth of its approximately 6.5 million acres to be slightly less than $3,175,000. Settlement of Pawnees, Poncas, Nez Percés, Otoes, and Missouris on 550,000 acres of Outlet land obligated the federal government, under provisions of the 1866 treaty, to pay the Cherokees more than $313,000. Payments made in 1880 and 1881 totaled more than $348,000, yet the National Council continued to instruct tribal delegates "to secure without further delay the remainder of the price still due for these lands." Under these urgings, the delegates asked Washington for an additional $500,000.[5]

Bushyhead attempted to analyze the problem in the spring of 1882. In an address to an extra session of the National Council, he described the Outlet as a region "steadily acquiring value" for grazing purposes and noted that "the profits or revenues to be derived therefrom should reach the highest equitable amount, with the greatest security." The Principal Chief had learned from sources in Washington that little likelihood existed "of other Indian tribes being removed" to the Outlet, "as the northern tribes justly protest against being sent thither." The land was thus available to the Cherokee Nation to do with as it willed—within reason, certainly—and Bushyhead believed that leasing might be the wisest use of this tribal resource. He suggested that arrangements be made with responsible parties "at a rate not less than two cents per acre . . . for not less than one year nor more than five years," with payments to be made semiannually. Lessees would be required to protect timber in the Outlet and to refrain from constructing permanent improvements. The mat-

5. H. Price to the Hon. Secretary of the Interior, December 30, 1884, U. S., Congress, Senate, 48th Cong., 2d sess., 1884, Executive Document 19, p. 2. An Act instructing and empowering the delegation to Washington, D. C. appointed under an act approved December A.D. 1881, CNP, WHC.

ter, he concluded, required immediate action.[6]

Cattlemen quickly learned of Bushyhead's views, but few of them showed any inclination to press the Cherokees for leases. Then, in December, two weeks after the Department of the Interior questioned their right to enclose Outlet range, several prominent ranchers, including Andrew Drumm and Benjamin S. Miller, wrote to the Principal Chief requesting leases on "the Ranges that have been both by courtesy and an unwritten law, considered as ours so long as the lawful taxes were paid." In an otherwise eloquent letter, they asked for protection from "citizens of the U. S. who steal the wood and bum the grass and who hold no grazing license" in the Outlet. "We trust that you will give this matter the consideration it merits," they concluded, "and us as the sincere friends of the Cherokees."[7]

The fencing controversy interrupted these informal lease negotiations. But while cattlemen fought attempts by the Department of the Interior to regulate their ranching operations, the United States Congress complicated matters by approving legislation to appropriate $300,000 "on account of the Cherokee lands lying west of the Arkansas River." The allocation would acquire greater significance later, when Senate committees tried to unravel the complicated question of Cherokee title to the Outlet. But in 1883 the congressional action, coinciding almost exactly as it did with the incorporation of the Cherokee Strip Live Stock Association, served to illustrate the extent of the federal government's preoccupation with the Outlet. The government had, in effect, made a substantial down payment on the land beyond the Arkansas, and on that basis it would later claim the region for the public domain.[8]

6. D. W. Bushyhead to the Honorable the National Council, May 2, 1882, File: Cherokee—Strip (Tahlequah), 1882, IAD, OHS.

7. A. Drumm, A. S. Raymond, et al. to Mr. D. W. Bushyhead, December 18, 1882, File: Cherokee—Strip (Tahlequah), 1882, IAD, OHS.

8. H. Price to the Hon. Secretary of the Interior, December 30, 1884, op. cit.

Against this background, the events of the spring and summer of 1883 unfolded in Tahlequah.

When Drumm and Eldred reached the Cherokee capital, they found leasing a recurring topic of conversation. Some Indians spoke of attempting to lease the Outlet themselves. If white men could graze cattle on a large scale beyond the Arkansas and realize substantial profits, so, they reasoned, could Cherokees. For years, Cherokee farmers had tried to run their few head of stock on the edge of the Outlet because of the dearth of adequate pasturage in the eastern portion of the Cherokee Nation, but federal troops had arrested and evicted them as intruders. Such treatment was unfair, they argued, as long as white men were permitted there. But the lure of the beef bonanza that swept the Great Plains in the 1880s was not lessened by such obstacles, and Cherokee stockmen felt as privileged as their white counterparts to reap the financial rewards of the range cattle industry. In any discussion of leasing, they believed, Cherokees should have priority on Cherokee land.[9]

Because of the divergence of opinion within the tribe on the lease question, the bicameral National Council created a joint committee of the council and senate to evaluate the situation in the Outlet and to suggest the form a new agreement with cattlemen might take. Cherokee stockmen requested that the committee consider their applications for a lease but later decided to air their views in open council. On May 10 the committee reported two lease bills, only one of which had originated in the council. Both named the Cherokee Strip

9. Testimony of James Madison Bell before the Senate Committee on Indian Affairs, January 28, 1885, Senate Report 1278, op. cit., I, pp. 262, 265.

Street scene, Tahlequah, in the 1880s

Live Stock Association as lessee. The council bill rented the Outlet to the association for five years at $100,000 per year; the second measure, which had not been regularly referred to the committee, stipulated that the association would receive a lease for a similar period if it matched the highest bid in excess of $100,000 per year offered for the Outlet by a competing group of cattlemen.[10]

On May 14 the first bill was read on the senate floor. Adhering closely to the suggestions offered by Bushyhead almost a year before, it forbade construction of permanent improvements in the Outlet and allowed "such temporary structures as may be absolutely required for the safe and profitable grazing of the stock" only; they were to be "held and declared . . . the property of the Cherokee Nation." According to the measure, cattlemen could not cut timber without permission from Cherokee authorities, and they were bound to prevent anyone else from doing so. They were not to obstruct public highways, stage lines, or mail routes. The bill also reserved for the Cherokees' sole use the major salt deposits in the Outlet.

The cattlemen's association, the bill stated, was to make payments of $50,000 at Tahlequah on October 1 and April 1 of each year. Grazing permits issued before passage of the bill would remain in effect until their expiration dates but would not be renewed. Ranchers who were not members of the association could not bring herds into the Outlet without permission from the association. Interlopers would be evicted by the Cherokees. Failure to meet payment dates or to comply with other conditions of occupancy would invalidate the lease. The Cherokee Nation, on the other hand, could terminate the agreement on six months' notice, should it decide to sell the Outlet. Should

10. Testimony of Robert Ross before the Senate Committee on Indian Affairs, May 22, 1885, Senate Report 1278, op. cit., II, p. 99. *The Cherokee Advocate* (Tahlequah), May 11 and 18, 1883.

Round-up and branding on the range

the association forfeit the lease, the Cherokees could negotiate a new one with other parties under the provisions of the same bill.[11]

On May 15 Senator Robert Ross, acting on behalf of the Cherokee stockmen, proposed an amendment to the bill that would change the words "Cherokee Strip Live Stock Association" to read "any responsible company of Cherokees." It was defeated. Ross then attempted to reduce the term of the lease from five years to three. That, too, was defeated. Another senator, perhaps acting on Bushyhead's instructions but in any case reflecting the Principal Chief's point of view, moved to raise the annual rental fee from $100,000 to two cents per acre, for an increase of $20,000 per year. The motion failed. The senate passed the bill by the required two-thirds majority, then sent it to the lower chamber. There, another attempt to shorten the five-year term miscarried. Although debate on the question delayed action, the deliberations required only a single afternoon. The bill was passed, returned to the senate, and sent to the Principal Chief. On May 19, Bushyhead approved it, and Drumm and Eldred formally notified him of their acceptance of its provisions.[12]

At Bushyhead's request, John F. Lyons drew up the lease. The Principal Chief left Tahlequah for Washington, to be absent for two weeks. He carried with him Lyons's draft for study. The attorney told Eldred that "if it suits him [Bushyhead], on his return, he will adopt it. I think it covers everything." [13]

Bushyhead evidently did not discuss the lease with either Teller or Price in Washington. Price had waited for three months after Teller's decision on enclosures without receiving any word from the Principal Chief. On June 28, however, he wrote to Tahlequah that

11. An Act to amend an act to tax stock grazing upon Cherokee lands west of the 96th meridian. U. S., Congress, Senate, 48th Cong., 2d sess., 1884, Executive Document 17, Part I, p. 151.

12. *The Cherokee Advocate* (Tahlequah), May 25, 1883. An Act to amend an act to tax stock grazing upon Cherokee lands west of the 96th meridian, op. cit., p. 152; A. Drumm and Chas. H. Eldred to Hon. D. W. Bushyhead, May 19, 1883, Senate Executive Document 54, op. cit., p. 156.

13. John F. Lyons to Charles H. Eldred, June 11, 1883, File: Cherokee Strip Live Stock Association (Section X), IAD, OHS.

unless the Office of Indian Affairs was notified of a new agreement between the tribe and Outlet cattlemen within twenty days, the order to remove barbed wire would be enforced.[14]

On July 5 Bushyhead and Charles Eldred, who held the power of attorney for the other association directors, signed the lease. It differed significantly from the enabling act that Bushyhead had approved a month before. Cattlemen were now permitted to cut Outlet timber for fuel, fence posts, "and other improvements as may be necessary and proper and convenient for the carrying on of their business." If the tribe sold the land, ranchers could remove whatever improvements they had constructed except those built with Cherokee timber. When the lease expired, or if it were forfeited, all improvements would become Cherokee property. In the event that the tribe sold only a portion of the Outlet, the association would receive on subsequent lease payments an annual rebate of 1.66 cents per acre for the land sold on subsequent lease payments.[15]

Three days later, the Principal Chief sent Price a copy of the May 19 act passed by the National Council. He emphasized that cattlemen's fences had become "the **property** *of the Cherokee Nation, as an attachment of the soil." Therefore, he said, "the main ground of complaint of such fencing, to wit, that its erection was an invasion of the rights of the nation, is . . . removed." Bushyhead erred. The act made fences tribal property, but the lease did not. Perhaps the Principal Chief suffered a momentary lapse of memory, but that seems unlikely, since he had spent nearly three weeks perusing Lyons's draft of the lease. Probably he was attempting to ease tensions between ranchers and the Department of the Interior. Clearly, he was the as-*

14. H. Price to Hon. D. W. Bushyhead, June 28, 1883, Senate Executive Document 54, op. cit., p. 155.

15. The Cherokee Strip Lease, Senate Executive Document 17, op. cit., p. 153.

sociation's friend. Its members had willingly paid grazing taxes for several years, and now, under the terms of the lease, they were prepared to pay more in a single year than Cherokee tax agents had been able to collect in four. Without hesitation Bushyhead would have exerted all the influence of his office to preserve the association's rights on Cherokee land, but, in obscuring the distinctions between the lease and its enabling legislation, he created a potentially troublesome situation. Had Teller or Price compared the two documents, the fencing controversy might not have been so easily settled.[16]

16. D. W. Bushyhead to Hon. H. Price, July 8, 1883, Senate Executive Document 54, op. cit., p. 156.

Although the association obtained its lease, opposition within the tribe to the cattlemen's occupation of the Outlet did not end. Cherokee stock raisers continued to grumble. Lyons, who remained in Tahlequah through the summer, had written Eldred in June that some Cherokees were "working hard against the lease" but as yet their efforts had "availed nothing." By August, criticism had grown more intense. The disgruntled Indian stockmen were joined by Elias C. Boudinot, the man who had helped to investigate conditions in the Outlet in 1879. By now an attorney residing in Washington, Boudinot was associated with homesteader groups and certain railroad interests that were committed to opening Indian Territory to settlement. He and a Cherokee named James Madison Bell had started a small ranching operation in the Outlet during 1882, but the entrenchment of white cattlemen there had ended the venture. Although Lyons knew of Boudinot's criticism of the lease, he did not consider it a serious threat to the association's occupancy of the Outlet. He acknowledged, however, that the lawyer was persistent, if nothing else. In November, Boudinot announced that he had been *"appointed to prosecute the Association," but his bombast still failed to impress Lyons.*[17]

The "lecturer," as Lyons called him, did not constitute the association's only opposition. Increasingly, Cherokees were becoming unhappy over their exclusion from the Outlet. Some tried to obtain grazing permits, even though the lease forbade further issuance of those documents. The situation prompted Lyons to ask Bushyhead for new assurances of his support for the association. The Principal Chief "desires me to say to you," Lyons wrote to Eldred, "that he entered into this whole matter in good faith and he intends to stand by it." Other Cherokees who claimed to have paid grazing taxes argued that they had thus acquired a right to the land. Bushyhead replied that taxes were assessed on the basis of livestock, not land, and that their claims had no validity. And as if this sparring were not enough to occupy the Principal Chief and the association's attorney, Lyons learned in September that some members of the National Council planned to introduce legislation approving the sale of the Outlet. Tribal opinion split evenly on the wisdom of such a move, so Lyons did not become alarmed. Then, in December, representatives of a Tennessee company arrived in Tahlequah and announced their willingness to buy for $1.25 per acre all the land currently leased by the association. Lyons told Eldred that the news did not "create much of a stampede," but at the same time he sought—and obtained—new assurances from Bushyhead that the tribe would not sell the land.[18]

Complaints from disappointed Cherokee stockmen and rumors about the sale of the Outlet were little more than minor irritants to the directors of the Cherokee Strip Live Stock Association. The cattle-

17. John F. Lyons to Charles H. Eldred, June 11, 1883, op. cit. Testimony of Elias C. Boudinot before the Senate Committee on Indian Affairs, January 9–10, 1885, Senate Report 1278, op. cit., I, pp. 100–101; Testimony of James Madison Bell before the Senate Committee on Indian Affairs, January 28, 1885, ibid., p. 265. John F. Lyons to Charles H. Eldred, August 15, 1883, File: Cherokee Strip Live Stock Association (Section X), IAD, OHS. John F. Lyons to Charles H. Eldred, November 23, 1883, ibid.

18. James Crutchfield to Hon. D. W. Bushyhead, June 30, 1883, File: Cherokee—Strip (Tahlequah), 1883, IAD, OHS. John F. Lyons to Charles H. Eldred, July 21, 1883, File: Cherokee Strip Live Stock Association (Section X), ibid. John F. Lyons to Charles H. Eldred, August 26, 1883, ibid. John F. Lyons to Charles H. Eldred, September 26, 1883, ibid. John F. Lyons to Charles H. Eldred, December 3, 1883, ibid.

men's position was relatively secure under the lease arrangement. They enjoyed the approval of the National Council and the Principal Chief, if not of the majority of the Cherokee people. Leaders in the tribal government saw the lease as mutually beneficial, and for that reason they supported the association's occupancy of the pasture beyond the Arkansas. Tahlequah's first concern was with tribal revenues, and chief and council alike willingly acted to realize profit from land that formerly had provided only limited benefits for the Cherokees. That the tribe reserved the right to terminate the lease whenever it decided to sell the Outlet is evidence of the fact that the Cherokees' commitment was not as firm as members of the association would have liked it to be. But the association represented a group of wealthy men, and few could be expected to challenge it on the grounds of capital resources. If the Outlet were to be sold, the most likely buyer would be the federal government—probably at a price well below the amount the association would be willing to pay for grazing privileges over an extended period of time. Thus, the relationship between the association and the Cherokee Nation acquired added significance. The Indians sought to derive revenue from their lands, and the ranchers wanted security for their businesses; the federal government could, however, prevent either group from achieving its goal. Cooperation between Cherokees and cattlemen was essential if such interference were to be avoided. But despite the grumblings that arose in 1883, the most serious challenges to the lease arrangement were yet to come.

IV. CERES AGAINST PALES

The most persistent opponents of the Cherokee Strip Live Stock Association were the homesteaders in Kansas, men and women on the trailing edge of the great wave of settlers that swept westward after the Civil War. As the fertile land of the Great Plains divided into quarter-sections beneath the surveyor's rod and yielded to the bite of the farmer's plow, latecomers looked toward Indian Territory for new homes. There they found that cattlemen were occupying land otherwise closed to settlement. Conflict was perhaps inevitable.

Elias C. Boudinot first stimulated the homesteaders' interest in Indian Territory with a series of letters written to newspapers early in 1879. In mid-February he revealed, in the Chicago **Times**, that the federal government had, after the Civil War, purchased millions of acres of land from tribes in the Territory. By treaties signed in 1866 with the Creeks, Seminoles, Choctaws, and Chickasaws, the government bought approximately 14 million acres for $1,600,530. More than 1 million acres had been assigned to Pottawatomies and Sacs and Foxes, while Wichitas held another 743,610 acres under an unratified agreement with Washington. The rest, Boudinot declared, was public domain. Located west of the 97th

Boomer leaders in 1884. ***Left to right:*** A. C. McCord, H. H. Stafford, David L. Payne, G. F. Goodrich, W. L. Couch, A. P. Lewis, Joe Pugsley

meridian and south of the Cherokee Outlet, this land was "well adapted for the production of corn, wheat and other cereals." [1]

Stimulated by Boudinot's report, prospective settlers flocked to the Kansas line. Secretary of the Interior Carl Schurz closely followed newspaper accounts of homesteaders' growing interest in Indian Territory and became alarmed by the situation. He announced that neither the Homestead Act nor any other federal land legislation applied to the purchased acreage. Any settlement in the area would be illegal. Schurz instructed the Commissioner of Indian Affairs to empower Indians to evict intruding farmers. Then, on April 26, 1879, President Rutherford B. Hayes issued a proclamation warning "certain evil-disposed persons" of the inadvisability of nesting on Indian land. [2]

In the spring of 1880, David L. Payne emerged as leader of the farmers who were "booming" for opening the Indian Territory to settlement. Payne, a onetime guide, scout, Kansas legislator, and petty bureaucrat, had met Boudinot in Washington. Both men enjoyed the support of railroads eager to remove barriers to homesteading below Kansas. Boudinot was content to seek lawful means of achieving those ends, but Payne preferred the more direct method of outright invasion. [3]

Between May 19, 1880, and August 28, 1882, the boomer leader was four times arrested within Indian Territory—once escorted to its borders and released, and three times jailed, either by Army or civilian authorities. His raids during 1883 occurred with such frequency that the War Department lost count of them. In the military's view,

> The whole history of Payne's operations is a farce, in which the Government is, of course, at a disadvantage. There is no punishment for Payne and his followers, the law only providing a fine for such

1. "Col. Boudinot's Letter, Showing the Status of the United States Lands in the Indian Territory," printed circular in Boomer Literature File, WHC.

2. "Unauthorized Settlement in the Indian Territory," printed circular in Boomer Literature File, WHC. Hayes repeated the warning again in 1880.

3. For origins of the Boomer movement, see Carl Coke Rister, *Land Hunger: David L. Payne and the Oklahoma Boomers*, chaps. 4 and 5; and Roy Gittinger, *The Formation of the State of Oklahoma, 1803–1906*, pp. 98–107.

> transactions—a sort of punishment easily borne by the impecunious crowd which follows this business of intrusion into the Indian Territory.[4]

Agent John Q. Tufts of the Union Agency at Muskogee shared the Army's attitude. At first, he had believed that Payne's evictions would convince the Indians of Washington's intention to protect their rights, but his conviction was short-lived. Payne's flagrant and repeated violations of the law brought heated protest from Tufts. Predicting that the Territory would be overrun by homesteaders, in the absence of substantial penalties for invasion, he characterized the government's efforts to exclude them as "a farce of the first water." Alert intruders could evade troops, but even if they were apprehended, the law's leniency allowed them to return almost immediately to the area from which they were evicted. Tufts saw governmental machinery breaking down under the weight of pointless arguments about which court held jurisdiction over Payne and his followers. "It makes little difference where they are tried," he wrote to the Commissioner of Indian Affairs,

> the result will be that they will be fined $1,000 each, and will inform the court that they are dead broke. . . . Payne and his crowd will be intruding again on the same land within six months. Until a law shall be enacted to punish by imprisonment for return to the reservation, after having been removed, it will be a physical impossibility to comply with the treaties to "remove and keep out all intruders" from an agency half as large as the State of New York, with a population of 100,000.[5]

Boomers turned their attention to the Cherokee Outlet in 1883. Boudinot had been careful to note that the Cherokee Nation had sold no territory to the federal government and that the Outlet was not public domain, but his delineation mattered little to homeseekers

4. Brief of papers showing action taken by the War Department in connection with invasion of the Indian Territory by D. I. [*sic*] Payne and others since April, 1879, U. S., Congress, Senate, 48th Cong., 2d sess., 1884, Executive Document 50, II, pp. 4–5. Hereafter cited as War Department Brief. Extracts from the annual reports of the commanding general, Department of the Missouri, relative to affairs in the Indian Territory in connection with the Oklahoma invasion, for the years 1879, 1880, 1881, 1882, 1883, 1884 [extract for 1883], ibid., p. 12.

5. John Q. Tufts to the Commissioner of Indian Affairs, September 30, 1881, *Annual Report of the Commissioner of Indian Affairs to the Secretary of the Interior for the Year 1881*, p. 104. John Q. Tufts to the Commissioner of Indian Affairs, August 29, 1884, *Annual Report of the Commissioner of Indian Affairs to the Secretary of the Interior for the Year 1884*, p. 99.

who had heard that there was "nothing finer in all the prairie world" than the "splendid piece of property" just below the Kansas line. Yet, while they coveted the land, their initial efforts to secure it proved futile because the settlers lacked an effective base of popular support. To create one, David Payne launched a three-pronged attack on ranchers in the Outlet. He began a propaganda campaign against them in the boomer newspaper, the ***Oklahoma War Chief;*** he argued for a congressional investigation of their lease arrangement with the Cherokee Nation; and he led settlers into their pasture, just as he had invaded land elsewhere in Indian Territory.[6]

Boomers hoped to attract public sympathy to their actions by pointing out that cattlemen occupied land from which all others were excluded. Leases of Indian domain, they argued, were sanctioned by the Department of the Interior in league with ranchers. They were mistaken. Secretary of the Interior Henry M. Teller's policy was one of nonrecognition, although he supported the rights of cattlemen who operated under leases against persons who held no agreement with the Indians. Teller's pronouncement, originally formulated with regard to leases in the Cheyenne–Arapaho reservation, was later applied to the Cherokee Outlet. The boomers contended that the lease arrangements promoted the growth of beef monopolies, postponed opening Indian land to settlement, and denied farmers access to vast areas of arable soil. "The Government," they stated, "owes to its own people every valuable tract of tillable land within our nation's borders." By approving the tribal leases, Washington ignored this obligation.[7]

Payne was obsessed by the idea that the cattlemen who used the Outlet wielded strong influence in the halls of Congress. Lavish dis-

6. *Oklahoma War Chief* (Wichita, Kansas), March 2, 1883.

7. H. M. Teller to Edward Fenlon, April 4, 1883, Senate, 48th Cong., 1st sess., 1883, Executive Document 54, IV, p. 99. *Oklahoma War Chief* (Wichita, Kansas), February 2, 1883.

tributions of cash in the proper places, he told farmers, in the prospect that they would become his followers, ensured federal protection for ranchers and financed the posting of troops in the field to hound homesteaders. Cattlemen had corporate connections and had "little trouble in getting the ear of the powers that be, at Washington." More than once Payne offered to prove publicly that ranchers had bribed Indians and federal officials to gain and maintain their lease.[8]

Payne's recurring sorties into Indian Territory soon split the homesteaders' ranks. Boudinot remained faithful to Payne, even to the point of formulating plans to file lawsuits against each rancher in the Outlet and declaring that the cattlemen had "violated the law ten times as much" as Payne. Others, however, questioned Payne's ability to lead. Some criticized his failure to stand fast in the face of federal opposition to his invasions; they encouraged him to "stay there [in Indian Territory] and not submit to any arrests or Escorts what ever." Still others viewed his repeated expeditions into the Territory as intolerable. They were responsible for raising money to finance his defense following each arrest, and the number and frequency of these demands on their pocketbooks were creating heavy expense. These supporters urged him to discontinue "unwarranted raids in the Territory." This conservative faction made its influence felt in August, 1884, when the Army again arrested Payne in the Cherokee Outlet. The weight of evidence was solidly against him, and homesteaders hesitated to have the matter adjudicated. Payne's lawyer, seeking a dismissal, promised cattlemen that the range in the Outlet would not "again be troubled in any way by the 'boomers'."[9]

Divisive elements within Payne's camp modified their differences

8. "To Our Oklahoma Colonists," leaflet in Boomer Literature File, WHC. D. L. Payne to Hon. Hiram Price, undated (received April 5, 1884), Letters Received, RG 75, RBIA, Special Case No. 111. This case contains identical letters to Secretary Teller, the Land Office, and the War Department.

9. B— to D. L. Payne, November 3, 1883, Miscellaneous Correspondence File for 1883, Payne Collection, OHS Library. "B—" is identified as Boudinot by a comparison of handwriting with that in Elias C. Boudinot to David L. Payne, April 27, 1882. Miscellaneous Correspondence File for 1882, ibid. John Hufbauer to Capt. D. L. Payne, August 31, 1883, Miscellaneous Correspondence File for 1883, ibid. "Your Friends" to Capt. D. L. Payne, September 20, 1883, ibid. J. Wade McDonald to Major Lyons, November 17, 1884, Miscellaneous Correspondence File for 1884, ibid.

Boomers under arrest by black troops. David Payne at center, with axe.

OKLAHOMA

CAPT. PAYNE'S

OKLAHOMA COLONY

Will move to and settle the Public Lands in the Indian Territory before the first day of December, 1880. Arrangements have been made with Railroads for

LOW RATES.

14,000,000 acres of the finest Agricultural and Grazing Lands in the world open for

FREE HOMES

For the people—these are the last desirable public lands remaining for settlement.

Situated between the 34th and 38th degrees of latitude, at the foot of Washita Mountains, we have the finest climate in the world, an abundance of water, timber and stone. Springs gush from every hill. The grass is green the year round. No flies or mosquitoes.

The Best Stock Country on Earth.

The Government purchased these lands from the Indians in 1866. Hon. J. O. Broadhead, Judges Jno. M. Krum and J. W. Phillips were appointed a committee by the citizens of St. Louis, and their legal opinion asked regarding the right of settlement, and they, after a thorough research, report the lands subject under the existing laws to Homestead and Pre-emption settlement.

Some three thousand have already joined the colony and will soon move in a body to Oklahoma, taking with them Saw Mills, Printing Presses, and all things required to build up a prosperous community. Schools and churches will be at once established. The Colony has laid off a city on the North Fork of the Canadian River, which will be the Capital of the State. In less than twelve months the railroads that are now built to the Territory line will reach Oklahoma City. Other towns and cities will spring up, and there was never such an opportunity offered to enterprising men.

MINERALS!

Copper and Lead are known to exist in large quantities—the same vein that is worked at Joplin Mines runs through the Territory to the Washita Mountains, and it will be found to be the richest lead and copper district in the Union. The Washita Mountains are known to contain Gold and Silver. The Indians have brought in fine specimens to the Forts, but they have never allowed the white men to prospect them. Parties that have attempted it have never returned.

In the early spring a prospecting party will organize to go into these Mountains, and it is believed they will be found rich in GOLD AND SILVER, Lead and Copper.

The winters are short and never severe, and will not interfere with the operations of the Colony. Farm work commences here early in February, and it is best that we should get on the grounds as early as possible, as the winter can be spent in building, opening lands and preparing for spring.

For full information and circulars and the time of starting, rates, &c., address,

T. D. CRADDOCK,
General Manager,
Wichita, Kan.

GEO. M. JACKSON,
General Agent,
506 Chestnut St., St. Louis.

October 23d, 1880.

Boomer handbill, 1880

to make common cause during the election campaigns of 1884. The Cherokee Strip Live Stock Association's president, Benjamin S. Miller, a candidate for the Kansas legislature, became the prime target of the boomers' concerted political enmity. Because of his connection with "that cattle monopoly," homesteaders believed that Miller would oppose opening the Indian Territory for settlement. He was, declared the ***Oklahoma War Chief***, "the natural financial and political enemy of the masses, and should be defeated." To prospective settlers, Miller represented "a soulless monopoly antagonistic to the rights of the people of the United States," an organization having "neither a legal or moral right" to use of the Outlet.[10]

Nine days after the election, but a week before its outcome was known, boomers assumed a more tolerant journalistic posture. "The CHIEF," an editorial in that paper announced, "makes no war on the cattle grazing or stock raising industry in the territories." As the world's principal supplier of beef, the United States was obliged to preserve as much land as was necessary for the range cattle industry. "Cattle is King," the newspaper conceded, "but in yielding fealty to our new Western monarch, let ALL THE SUBJECTS have an equal chance to make States and build cities where now is solitude." Within a week, however, homesteaders forgot the monarch to praise the messiah. Believing that Grover Cleveland's election ensured the opening of the Outlet and the purchased land known as the Oklahoma District, boomers gleefully chanted, "SQUAT, SQUATTERS, SQUAT." But the celebration was premature.[11]

10. *Oklahoma War Chief* (South Haven, Kansas), October 23 and 30, 1884.

11. *Oklahoma War Chief*, November 13 and 20, 1884.

In their sparring with boomers, cattlemen in the Outlet were not without effective allies. Newspapers in Caldwell, Kansas, had criticized Payne's tactics even before the homesteaders had drawn support from or were represented in the fourth estate. The **Caldwell Post,** founded in 1879, was editorially undistinguished and given largely to publishing tasteless lampoons of Payne and his followers until Tell W. Walton became its editor in 1881. J. D. Kelly, Jr., had edited the paper when it first appeared on January 2, 1879, and J. S. Sain assumed control in December, 1880. The new editor was a man of some standing among local cattle interests, and under his direction the paper grew in influence. A rival paper, the **Caldwell Commercial,** appeared in May, 1880, under the editorship of William B. Hutchison, a veteran Wichita newspaperman and a charter officer of Payne's boomer organization. That affiliation notwithstanding, Hutchison assumed a moderate position on the question of opening the Indian Territory to settlement, but gradually his editorial policy shifted away from the homesteaders' cause. Perhaps he was wooed away by the Outlet ranchers' money. Cattlemen's range and brand advertisements began to appear with increasing regularity during the period of the **Commercial**'s editorial transformation.[12]

In May, 1883, the **Post** and the **Commercial** merged under the control of the Caldwell Printing and Publishing Company, a joint-stock venture headed by Benjamin S. Miller and John W. Nyce. Thereafter, the company published only one paper, the **Caldwell Journal,** edited by Hutchison. It was solely a cattlemen's newspaper, supported almost exclusively by advertising from Outlet ranchers. On October 3, 1883, the Cherokee Strip Live Stock Association an-

12. *Caldwell Post*, December 23, 1880. Compare *Caldwell Commercial*, May 13 and 17, 1880, with issues for December 14, 1882, and February 1 and 15, 1883.

nounced that henceforth the **Journal** would be its "official paper." [13]

If Outlet cattlemen found it advantageous to buy their own newspaper for rhetorical exchanges with homesteaders, they also gained immeasurable satisfaction from the journalistic endeavors of their staunchest ally, the Cherokee Nation. The **Cherokee Advocate,** official voice of the Indian government at Tahlequah, entered the fray early to defend the rights of taxpaying and, later, leasing cattlemen against those of prospective settlers. The policies of the Cherokee press were not influenced by the pittance to be derived from advertising, for a larger sum was at stake. Accordingly, interlopers on the tribal lands—whether renegade cattlemen or homesteaders intent upon opening the country—were treated alike. On the other hand, ranchers who, through the Cherokee Strip Live Stock Association, met their obligations as tenants were considered friends, standing "squarely with the Cherokees to defend the rights of both [Indian and stockraisers] against the lawless class." [14]

The association's members readily recognized the economy of the alliance, even though it did not always function as smoothly as they thought it should. Under the lease provisions, the Cherokees were assigned the task of evicting intruders from the Outlet, and as boomer activity increased, cattlemen called more frequently on Cherokee officials at Tahlequah for help in policing their range. Initially, each party to the lease assigned this burden to the other, but continuing incidents of intrusion led Indians and ranchers alike to devise new methods of dealing with their mutual problems. Tests of their effectiveness were to be heavily significant in the futures of both the association and the tribe.

13. *Caldwell Post*, April 26 and May 3, 1883; *Caldwell Journal*, May 17 and 24 and October 4, 1883.

14. *Cherokee Advocate* (Tahlequah), October 3, 1884.

V. ROCK FALLS AND CHILOCCO

On August 23, 1883, Augustus E. Ivey, a citizen of the Cherokee Nation and sometime journalist who resided in Vinita, wrote to Secretary of the Interior Henry M. Teller, charging that the Cherokee Strip Live Stock Association had acquired its lease of the Outlet "through the most corrupt means." Ivey had once grazed a few head of stock west of the Arkansas, but the lease arrangement now denied him access to the range. He believed that many other Cherokees had also been evicted by—as he termed it—the cattlemen's robbery of the Nation. "Could the inside of the scheme be seen through," he wrote, "—and it can—I dare say no more vile a swindle was ever perpetrated upon our people." [1]

Ivey's letter circulated in Washington but prompted no action for more than a year. Eventually it reached the desk of Senator George G. Vest of Missouri. On December 2, 1884, Vest informed Henry L. Dawes, Chairman of the Senate Committee on Indian Affairs, that he could provide "names, amounts, and dates, which show that as widespread a scheme of corruption is today in existence in that Indian Territory as ever obtained in the worst times and under the worst methods known to the States . . . or any other community." Vest's comments spread quickly, and within twenty-four hours the Senate passed a resolution instructing Dawes's committee

1. Augustus E. Ivey to the Secretary of the Interior, August 23, 1883, U. S., Congress, Senate, 48th Cong., 1st sess., 1883, Executive Document 54, IV, p. 160.

to determine the extent to which leases had been made in Indian Territory and to compile lists of the signatories. The committee was to investigate the process by which cattlemen acquired leases and to decide whether or not such agreements were "conducive to the welfare of the Indians." [2]

2. U. S., *Congressional Record*, 48th Cong., 2d sess., 1884, XVI, Part I, p. 11. U. S., Congress, Senate, 49th Cong., 1st sess., 1885, Senate Report 1278, Part I, p. i.

This, then, was the situation that cattlemen and Cherokees faced in 1884. Confronted on the one hand by the threat of a congressional investigation and on the other by the reality of increasing invasions by boomers, they struggled to defend their rights. Success required cooperation, and that was born of their desperation.

Late in the spring of 1884, Benjamin Miller notified Chief Bushyhead that intruders were in the Outlet below Hunnewell, Kansas. Other cattlemen reported the invasion, adding the information that their fences had been cut by homesteaders crossing their range. Some suggested that the Cherokees make the eviction of the settlers an opportunity to eject not only the farmers but also stockraisers who paid no association dues. Bushyhead responded by sending Miller a bundle of no-trespassing notices for distribution among homesteaders and unwelcome stockmen. Miller's response was heated: "If, at the present time, I should serve notices upon each individual intruder on lands leased by our Association from the Cherokee Nation, I should have to serve more than a thousand and it would be rather arduous." [3]

3. Ben S. Miller to D. W. Bushyhead, May 31, 1884, CNP, WHC. Roberts and Windsor to D. W. Bushyhead, June 4, 1884 (copy), ibid. Ben S. Miller to D. W. Bushyhead, June 23, 1884, ibid.

Reports of invasion continued to reach Bushyhead, some setting the number of boomers in the Outlet as high as two thousand. When

Payne and his colony laid out a town, which they named Rock Falls, on the banks of the Chikaskia River in the Outlet south of Hunnewell and set up headquarters for the **Oklahoma War Chief** *there, a director of the association, E. M. Hewins, wrote of "Cheeky fellows [who moved] a printing office on Cherokee lands to demean them and lye about [the Indians] in [a] most Shameful Manner." Cherokee officials, acting through Agent Tufts, had already advised officials in Washington of the situation; on July 1, 1884, President Chester A. Arthur, in a proclamation similar to Hayes's, warned homesteaders away from Indian Territory, declaring that invaders would be "speedily and immediately removed . . . by the proper offices of the Interior Department [with, if necessary,] the aid and assistance of the military forces of the United States." Payne ignored the warning.*[4]

The responsibility of removing boomers from Rock Falls fell to John Q. Tufts. On July 22 he assigned Connell Rogers, a Union Agency clerk, to represent the Bureau of Indian Affairs in the matter. On July 23 Rogers appeared at Payne's camp and presented an order to move. Payne refused; he produced a map indicating that the Cherokee Outlet had been ceded to the government on July 19, 1866, and was therefore public domain. Rogers, seeing that he could not reason with the boomers' leader, withdrew and requested military assistance.[5]

On the evening of August 6, three men rode into Rock Falls. They were Col. Edward Hatch, commander of the Military District of Oklahoma, Department of the Missouri; Lt. W. Leighton Finley, the district's acting assistant adjutant-general; and A. R. Greene, an inspector from the General Land Office. They formally notified Payne

4. E. M. Hewins to D. W. Bushyhead, July 6, 1884, CNP, WHC. War Department Brief, p. 14.

5. John Q. Tufts to the Commissioner of Indian Affairs, September 9, 1884, Letters Received, RG 75, Special Case No. 111, RBIA. Connell Rogers to Col. John Q. Tufts, August 18, 1884 (copy), ibid.

Connell Rogers, representative of the
Union Agency in the raid on Rock Falls

Col. Edward Hatch

Boomer camp near Caldwell, Kansas, summer of 1885

and his followers to leave the Outlet quickly and quietly or face arrest by federal troops. Payne again refused to move. Instead of complying, he threatened to arrest Colonel Hatch. The die was cast.[6]

At ten o'clock on the morning of August 7, Rogers, Inspector Greene, and John F. Lyons arrived at the boomer settlement with Companies L and M of the Ninth United States Cavalry, commanded by Capt. Francis Moore. The troops placed the homeseekers under arrest and escorted them to the Kansas border, where they were released and warned not to return. Payne and several of his lieutenants were taken to Fort Smith, Arkansas, for trial in United States district court as "old offenders under the law." The arrests completed, the soldiers confiscated the boomers' printing press and burned Rock Falls to the ground.[7]

John F. Lyons's presence at Rock Falls underscored his importance in maintaining close liaison between the association and the Cherokee government. In addition to representing the association's leaders at the raid, Lyons reported the incident in minute detail to Chief Bushyhead. Lyons's activities suggest that the cattlemen and the Indians were developing informal machinery for dealing with problems connected with the Outlet—machinery that could, if necessary, circumvent procedures established by the federal bureaucracy. The Rock Falls raid had been planned and executed according to a rigid protocol: the boomers intruded, the cattlemen notified the Cherokees, and—after some confusion—the Cherokees informed the Department of the Interior through Agent Tufts. Nevertheless, neither Indians nor ranchers relinquished their avenues of communication with each other. The cooperation revealed by the raid, however rudimentary,

6. Rogers to Tufts, op. cit. War Department Brief, p. 5. William H. Leckie, *The Buffalo Soldiers: A Narrative of the Negro Cavalry in the West*, pp. 7, 245; *Annual Report of the Commissioner of the General Land Office for the Year 1885*, p. 50.

7. Rogers to Tufts, op. cit. Payne was no stranger to that court. See Ed Bearss and Arrell M. Gibson, *Fort Smith: Little Gibraltar on the Arkansas*, pp. 326–27. For a full account of the incident, see William W. Savage, Jr., "The Rock Falls Raid: An Analysis of the Documentary Evidence," *The Chronicles of Oklahoma*, 49:1 (Spring 1971), 75–82.

reflected an important development, one that would become fully apparent as the result of another incident in another part of the Outlet.[8]

8. John F. Lyons to D. W. Bushyhead, November 19, 1884 (copy), File: Cherokee—Strip (Tahlequah), 1884, IAD, OHS.

In January, 1884, the federal government opened an Indian school on the banks of Chilocco Creek in the Outlet, six miles south of Arkansas City, Kansas. The precedent of settling friendly tribes in the Outlet established by the 1866 treaty provided the basis for the school, which was authorized by an Indian appropriation act passed by Congress on May 17, 1882. Approximately 1,200 acres of Outlet land were designated for the purpose, but when the buildings were completed, officials concluded that the acreage was inadequate to sustain the extensive agricultural program planned for the school. They therefore recommended enlarging the site. In a proclamation dated July 12, 1884, President Chester A. Arthur added 7,440 acres to the Chilocco lands, taking for government use about fifteen sections of pasture that had been subleased from the association by the firm of Roberts and Windsor.[9]

9. Charles J. Kappler, comp. and ed., *Indian Affairs: Laws and Treaties,* I, 493. For the origins and development of the school, see Larry L. Bradfield, "A History of Chilocco Indian School," Master's thesis, University of Oklahoma, 1963.

Roberts and Windsor retained an attorney, W. P. Hackney of Winfield, Kansas, and together they besieged Bushyhead with requests for assistance. The naive Hackney sought to determine "upon what the President bases his authority" in order to challenge it. His employers, however, pursued a more practical course. Arguing that the location of the new school boundaries left them without access to water on one corner of their range, Roberts and Windsor suggested that Bushyhead exert his influence to have the lines redrawn. But the

Principal Chief recognized that the situation was beyond his control. The association's officers said nothing, and there matters rested for nearly one year. When problems next arose, school officials, not cattlemen, were the complainants.[10]

On March 26, 1885, Henry J. Minthorn, superintendent of Chilocco, wrote to Commissioner of Indian Affairs Hiram Price to protest a series of intrusions on school lands by a group of unidentified cowboys. Minthorn, a Republican, had tendered his resignation from the school on March 4, perhaps anticipating a Democratic house-cleaning after the inauguration of President Grover Cleveland, but he had not been notified of its acceptance. The spoils were already being divided in Washington, and Minthorn's letter reached J. D. C. Atkins, who had replaced Price as commissioner in the Indian Bureau.

Minthorn complained that two herds of cattle had been driven into plowed fields near the school, reducing the students' chances of raising a crop and threatening the school's cattle with contagious disease. He had sought assistance in removing the cattle at a nearby Army encampment, but was refused. Boomers had "some just reason to complain," he said, when federal troops held them outside of an area open to ranching. "From what I have seen of cattle men," Minthorn wrote the Commissioner, "I should say they are the last class of men that should be admitted to this Territory." [11]

Atkins sent Minthorn's letter to Secretary of the Interior L. Q. C. Lamar, remarking with the zeal of a new appointee that "justice and sound policy require that cattle herds shall be removed **without delay** from the vicinity of the tract set aside for the Chilocco school."

10. W. P. Hackney to Hon. D. W. Bushyhead, July 28, 1884, File: Cherokee—Strip (Tahlequah), 1884, IAD, OHS. Roberts and Windsor to Hon. D. W. Bushyhead, November 25, 1884, ibid.

11. H. J. Minthorn to Hon. H. Price, March 26, 1885, File: Cherokee—Strip (Tahlequah), 1885, IAD, OHS.

Lamar notified Bushyhead of the situation and advised him to "take early measures to prevent intrusion . . . upon the school lands [so] there will be no further occasion for complaints of this character." [12]

On April 19 Bushyhead ordered John W. Jordan to investigate the alleged intrusions. Jordan, a Cherokee farmer and stockman, had written to the Principal Chief early in January to express concern over conditions in the Outlet. He urged Bushyhead to prevent the erosion of Cherokee claims to the land by reaffirming the tribe's right of possession and jurisdiction over it. The boomer problem was uppermost in Jordan's mind, as it was in Bushyhead's when, on February 1, he appointed Jordan to act as the Nation's special agent for the Outlet. Yet, Jordan's first assignment apparently concerned cattlemen rather than homeseekers.[13]

Jordan rode to Arkansas City in May with the association's secretary, John A. Blair. There they investigated the Chilocco intrusions and soon concluded that no members of the association were involved. "The men that were there," Blair later told Bushyhead, "are from the states with small bunches of cattle." Jordan had little sympathy for either the school or the problems Minthorn had described and placed blame for the incident on the school's officials. The outspoken Minthorn may have antagonized him during the investigation, but in any case, Jordan dismissed the episode and did not mention Chilocco in his annual report to Bushyhead.[14]

The Rock Falls raid and the incident at Chilocco were significant for several reasons. First, they revealed the concern of cattlemen in the association for maintaining a favorable corporate image in the eyes of the federal government. The association's reluctance to argue

12. J. D. C. Atkins to the Secretary of the Interior, April 10, 1885, File: Cherokee—Strip (Tahlequah), 1885, IAD, OHS. L. Q. C. Lamar to Hon. D. W. Bushyhead, April 11, 1885, ibid.

13. J. W. Jordan to Hon. D. W. Bushyhead, April 30, 1885, File: Cherokee—Strip (Tahlequah), 1885, IAD, OHS. D. W. Bushyhead to Mr. John Jordan, February 1, 1885, ibid.

14. John A. Blair to Hon. D. W. Bushyhead, May [n.d.], 1885, File: Cherokee—Strip (Tahlequah), 1885, IAD, OHS. J. W. Jordan to Hon. D. W. Bushyhead, April 30, 1885, ibid. John W. Jordan to The Honorable The Principal Chief, November 6, 1885, ibid.

John W. Jordan, special agent, Cherokee Nation

the matter of the school's location and the willingness with which its representatives assisted Cherokee officials in investigating Minthorn's complaint indicated its determination to avoid conflict at a time when Congress was focusing attention on the question of white men's leases of Indian land. The fact that cattlemen acknowledged the complaint at all was a strong indication of their estimate of the damage it could have done them. The same realization governed their adherence to federal procedure in removing intruders from Rock Falls.

Both incidents reflected the growing reliance that the association and the federal government placed on the Cherokee Nation whenever problems arose that concerned the Outlet. Washington was quick to supervise, to the extent of formulating policy, but it was often slow to act when those policies required enforcement. Owing perhaps to the complexity of bureaucratic processes that in many cases impeded rather than facilitated the solution of problems, federal officials sought to confer upon Cherokees a limited police power. Ordinarily, requests for military assistance to remove intruders that originated in the Department of the Interior required approval of the Secretary of War. Once the War Department issued orders, action by military district and department commanders was required, followed at last by coordination of troops in the field by post commanders. The procedure was time-consuming. It could be bypassed, however, if the Department of the Interior maintained direct channels of communication with the Cherokee government at Tahlequah beyond those provided by the Union Agency. Special agents like John W. Jordan were certainly less effective than a detachment of cavalry, but they had the advantage of being immediately available. The matter of the Chilocco intrusions was settled quickly, while similar occurrences involving boomers in which federal troops were used required more time.

If Washington depended on the Cherokees for the sake of expediency, the Outlet cattlemen relied on them for reasons of simple convenience. Despite the structure of their association, which provided both the money and manpower necessary to organize an adequate protective force, the ranchers were hesitant to police their own ranges. The instability of their lease arrangement made them sensitive to government criticism. Federal agencies had established procedures, and the cattlemen had no desire to incur their wrath by working at cross-purposes with them. On the other hand, the Cherokee Nation was not a part of the federal bureaucracy and could be more easily approached by the association. Moreover, Outlet ranchers, as businessmen during the peak years of the beef bonanza, were too involved with their entrepreneurial activities to divert men and equipment for range patrols; it should be recalled that they had abandoned a similar practice years earlier because it had been too expensive. Further, through their association the cattlemen paid the Cherokees a substantial annual rental for range in the Outlet, and they expected their rights to be protected.

The Chilocco episode provided the Cherokees and the cattlemen with an opportunity to test the machinery they had devised for solving problems in the Outlet. If the cattlemen feared federal displeasure with the lease arrangement as a threat to their economic well-being, the Cherokees viewed it as a real menace to their national sovereignty.

They sought at once to derive revenue by leasing land to the association and to preserve, vis-à-vis the federal government, their right of possession and jurisdiction with regard to the Outlet. Thus, they were willing to act on the cattlemen's complaints and to assist the Department of the Interior in investigating reports like Minthorn's—unless the interests of the association and the federal government were in conflict. That possibility was always to be considered, and when such situations arose, the government's interests took precedence. The Cherokees accepted the government's sanctioning of their police power, but they wisely recognized the limitations of that power. The machinery worked within specified areas and with fairly obvious restrictions, although few other than tribal and association leaders understood its functioning.[15]

Chief Bushyhead, in all matters involving cattlemen and the federal government, was in effect a man in the middle. Discussing his responsibilities, he once wrote:

> Empowered, as he is, to see the Constitution [of the Cherokee Nation] observed and the laws faithfully executed, it is the Chief's duty to extend the supervision of this [Executive] Dept. over every part of the Cherokee domain, as far as possible with the means at his command. . . . It is his duty to see that the law . . . be faithfully executed, and, in accordance therewith, to prevent intrusions, and have intruders expelled . . . and to see that the common property . . . is protected against depredations by any party.[16]

On the one hand, he could appoint or receive agents like Jordan and Lyons to assist him and to act as liaison between his office and the association; on the other, he frequently had to mediate personally

15. For an indication that even Jordan occasionally failed to understand the reason for Bushyhead's guidance of his activities, see J. W. Jordan to the Hon. Cherokee Chief D. W. Bushyhead, November 1, 1886, File: Cherokee—Strip (Tahlequah), 1886, IAD, OHS.

16. D. W. Bushyhead to Mr. John Jordan, February 1, 1885, op. cit.

when problems arose, and, more importantly, he was obligated to represent the best interests of his people when confronting those departments of government with whom ultimate control of the tribe's destiny rested. It was no easy task, but it was one to which Bushyhead was equal. For example, if the Chilocco affair amounted to nothing in terms of subsequent events, the credit for keeping it so must be largely his. In the beginning, certainly, it was potentially damaging to both Cherokee sovereignty and the association's tenancy. Seen in that light, its resolution was a significant achievement.

The Chilocco incident was but a brief, diverting current in the tide of boomer activity. David Payne's death in November, 1884, did little to ease tensions between homeseekers and cattlemen, and when leadership of the homesteaders' movement fell to W. L. Couch, it became apparent that there would be no attempt by boomers toward conciliation with the cattlemen. The boomers' newspaper dropped its bellicose title and became simply ***The Oklahoma Chief***, a change that belied its hardening attitude toward Outlet ranchers. "We shall oppose," it stated, "all monopoly by syndicates and cattle rings of the public domain to the exclusion of the homestead settlers." There could be no question about Couch's intentions.[17]

In January, 1885, Couch, already branded a fanatic in Army reports, led a party of four hundred settlers across the Outlet and into the Oklahoma District, encamping near Stillwater. The boomers remained there until elements of the Ninth Cavalry cut supply lines

17. *The Oklahoma Chief* (Arkansas City, Kansas), February 3, 1885.

and forced them to surrender. The homesteaders' resistance perhaps nudged the pendulum of public opinion slightly toward their side, and it almost certainly impressed many legislators in the United States Congress. In March, President Cleveland signed legislation authorizing government negotiations for the purchase of the Oklahoma District and part of the Cherokee Outlet. That authorization marked what was in effect the high point of the boomer movement, for, as Roy Gittinger has observed, "Boomers henceforth had only to see that the government did not neglect to press the matter to a successful conclusion." The Cherokee Strip Live Stock Association faced challenge once again.[18]

18. War Department Brief, pp. 7–8. Origins of the legislation are traced in Roy Gittinger, *Formation of the State of Oklahoma, 1803–1906*, pp. 127–30.

Generations of Americans—historians among them—who have been reared on the casual violence in the works of Zane Grey and Max Brand have accepted unquestioningly the stereotype Western showdown between cattlemen and homesteaders—steel-eyed antagonists who gave no quarter and asked for none, men who lived close to death and found salvation only in bloodshed. That image is today constantly reinforced by mass media. In the popular mind, mention of farmer–rancher confrontations in a Western context triggers sanguinary visions of fence-cutting raids, burning homesteads, and blazing gunfights in a dry riverbed, symbolic gateway to the promised land or a quarter-section thereof. Events in the Cherokee Outlet—of which the Rock Falls raid and the Chilocco incident were typical—must stand as a correction to this concept. There, the encounter between rivals, whether cattlemen and homeseekers or

Indians and drovers, was bloodless and the violence rhetorical.[19]

Some writers have suggested that historians begin "to differentiate between rhetoric that acts as a benign substitute for extreme action and rhetoric that is only a prelude to such action" when considering verbal or ideological violence. It is possible with some certainty to place in the first category the hostility that is apparent in newspapers and correspondence dealing with farmer–rancher feuds over the Outlet. Many threats were made, but few were carried out. The raid on Rock Falls—destruction of property aside—was in the main a face-to-face exchange of invective. The heavily armed participants spilled no blood that day.[20]

The frontier of the Outlet cattleman and the Kansas homesteader was agricultural. Whether or not one accepts the sequence of Turner's "procession of civilization," the agricultural frontier was certainly a terminal phase of the westering process. When the farmer arrived at the edge of Indian Territory, he found it ringed by the forms, if not the substance, of civilization. The law, in both its military and civilian guises, was everywhere apparent. Against that background the boomer began his campaign to open the Territory, and within that framework the cattleman worked to preserve his last stronghold. These circumstances, centering around the availability of legal recourse, may have accounted for the absence of physical violence. And if the events in the Outlet serve as a correction—however mild—to the farmer–rancher stereotype, it is also further evidence of the uniqueness of the association and its milieu.[21]

19. Joe B. Frantz, "The Frontier Tradition: An Invitation to Violence," in Hugh David Graham and Ted Robert Gurr, eds., *Violence in America: Historical and Comparative Perspectives*, pp. 127–54. The origins of the theme in popular literature are traced in Russel Nye, *The Unembarrassed Muse: The Popular Arts in America*, pp. 280–304. Robert Easton, *Max Brand: The Big "Westerner."* The historical background commonly said to support such distortions is generalized in Joe B. Frantz and Julian Ernest Choate, Jr., *The American Cowboy: The Myth & the Reality*, pp. 110–11.

20. Richard Hofstadter and Michael Wallace, eds., *American Violence: A Documentary History*, p. 5. Statement by H. E. Horn in T. N. Athey, "Historical Biography of David L. Payne," manuscript, Athey Collection, OHS, pp. 28–29. W. Merritt, Brigadier General Comdg., Headquarters Department of the Missouri, to Assistant Adjutant General, Division of the Missouri, August 8, 1888 (copy), Letters Received, RG 75, RBIA. Savage, "The Rock Falls Raid," pp. 81–82.

21. Frederick Jackson Turner, "The Significance of the Frontier in American History," in American Historical Association, *Annual Report of the American Historical Association for the Year 1893*, p. 208.

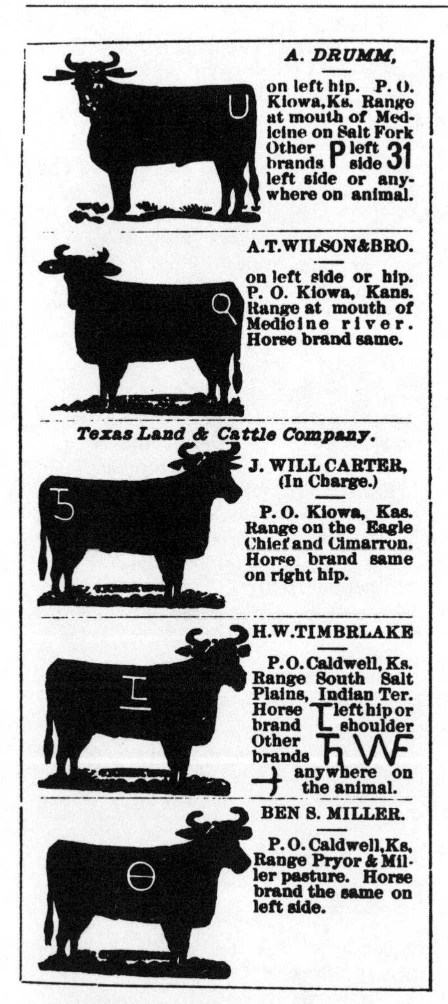
A. DRUMM,

on left hip. P. O. Kiowa, Ks. Range at mouth of Medicine on Salt Fork Other brands P left side 31 left side or anywhere on animal.

A.T.WILSON&BRO.

on left side or hip. P. O. Kiowa, Kans. Range at mouth of Medicine river. Horse brand same.

Texas Land & Cattle Company.

J. WILL CARTER, (In Charge.)

P. O. Kiowa, Kas. Range on the Eagle Chief and Cimarron. Horse brand same on right hip.

H.W.TIMBRLAKE

P.O. Caldwell, Ks. Range South Salt Plains, Indian Ter. Horse brand left hip or shoulder Other brands HWF anywhere on the animal.

BEN S. MILLER.

P.O. Caldwell, Ks, Range Pryor & Miller pasture. Horse brand the same on left side.

Cattle brands from ***South-Western Brand Book*** for 1883

VI. CATTLEMEN BESIEGED

The Senate Committee on Indian Affairs met on December 9, 1884, to begin inquiries into the association's lease as outlined by Augustus Ivey in his letter to Secretary Teller, dated August 23, 1883. The committee's task was complicated by another Senate resolution that assigned to it the additional work of investigating the status of freedmen in Indian Territory, the relationship between the various tribes and the federal government, the general condition of the tribes, and problems concerning reservation boundaries. As if these duties were not enough to occupy the members' attention, the committee was also to examine the need for new federal legislation affecting Indian policy.[1]

Hearings began in Washington in December, 1884, and continued through the summer of 1885, when the committee traveled to Kansas and Indian Territory to interrogate witnesses who were unable to journey to Washington. The senators' questions covered the spectrum of items included in their instructions, but the bulk of their work concerned the Cherokee Strip Live Stock Association lease. Ranchers, Cherokees, homeseekers, and federal officials were called to testify.

1. U. S., Congress, Senate, 49th Cong., 1st sess., 1885, Senate Report 1278, VIII, Part I, pp. i, 3.

Of the ten Outlet cattlemen who appeared before the committee, eight had been members of the original board of directors of the Cherokee Strip Live Stock Association: Benjamin S. Miller, S. Tuttle, Andrew Drumm, James W. Hamilton, Charles H. Eldred, A. J. Day, M. H. Bennett, and E. M. Hewins. The testimony of Bennett and Hewins did not concern the Outlet lease but dealt, rather, with charges that their stock had drifted south into the Oklahoma District. The combined range subleased from the association by these men totaled nearly 1 million acres. The other two cattlemen were the association's secretary and a rancher who held an interest in 400,000 acres of leased land. All the cattlemen argued that the association had done nothing unlawful in securing the lease from the Cherokee Nation, and most of their testimony merely chronicled the origins and growth of their organization and described its administrative machinery.[2]

W. L. Couch presented the homesteaders' view of the lease. When the boomers' leader learned of the committee's investigation, he wrote to Dawes, asking to be allowed to testify. He offered to present evidence of bribery by Outlet ranchers and claimed that some government officials had accepted money from cattlemen in return for special consideration in Congress. The committee, meeting in Caldwell in June, 1885, invited Couch to appear before it.[3]

Couch charged that ranchers whose range in the Outlet bordered on the Oklahoma District often wintered stock across the line on land denied the homesteaders. But that point was not viewed as central to the issue, so the senators pressed Couch to elaborate on the allegations he had made in his letter to Dawes. Thereupon, Couch explained to the committee that the association's lease was invalid

2. Testimony of Benjamin S. Miller, January 9, 1885, Senate Report 1278, op. cit., pp. 79 ff; Testimony of S. Tuttle, January 9, 1885, ibid., pp. 89 ff; Testimony of Andrew Drumm, January 9, 1885, ibid., pp. 75 ff; Testimony of James W. Hamilton, January 9, 1885, ibid., pp. 91 ff; Testimony of Charles H. Eldred, January 13, 1885, ibid., pp. 149 ff; Testimony of M. H. Bennett, June 6, 1885, ibid., Part II, pp. 424 ff; and Testimony of E. M. Hewins, June 6, 1885, ibid., pp. 433 ff. Statement of the secretary of the Cherokee Strip Live Stock Association, ibid., Part I, pp. 308–9. Testimony of J. A. Blair, January 21, 1885, ibid., p. 180; Testimony of H. L. Newman, December 9, 1884, ibid., p. 33.

3. W. L. Couch to Hon. Henry L. Dawes, June 4, 1885, Senate Report 1278, op. cit., Part II, Appendix, pp. 18–19.

because Cherokees could not lease what was not theirs. The Outlet, he said, was the property of the federal government, acquired from the Cherokee Nation by the 1866 treaty. Asked if he had read the treaty, Couch replied, "Yes . . . that and acts of Congress." Asked to identify the acts to which he referred, Couch answered, "I don't know that I can cite them particularly." When Dawes questioned Couch closely about his charges of bribery, the boomers' leader admitted that his evidence was secondhand. Statements made to him by members of the Cherokee Strip Live Stock Association had led him to believe that the cattlemen had spent $36,000 to secure the lease and to obtain approval for it from the Department of the Interior.[4]

John Q. Tufts of the Union Agency maintained before the committee that the Department of the Interior allowed the lease because it "recognized that these Indians have the right to transact their own business without our interference." He told the senators that he saw no need for their investigation. He expressed his belief that the lease arrangement was fair to the Cherokees; he knew of nothing to indicate that the cattlemen had bribed the Indians.[5]

Citizens of the Cherokee Nation comprised the largest single group to appear before the committee. Of the twenty-two Cherokee witnesses, twelve opposed the association's lease and ten defended it. Predictably, perhaps, seven of the ten advocates were connected in some way with the Cherokee government. Another was a white man, a Cherokee by adoption.[6]

The Cherokee Strip Live Stock Association, through its board of directors, attempted to discredit the testimony of several Cherokees who criticized the lease. The association's attorney John F. Lyons

4. Testimony of W. L. Couch, June 6, 1885, Senate Report 1278, op. cit., pp. 445–47, 451, 457.

5. Testimony of John Q. Tufts, January 6, 1885, Senate Report 1278, op. cit., Part I, pp. 28, 30.

6. Details of the Cherokee testimony are in William W. Savage, Jr., "Leasing the Cherokee Outlet: An Analysis of Indian Reaction, 1884–1885," *The Chronicles of Oklahoma*, 46:3 (Autumn 1968), 285–92.

Residence of William F. Rasmus, Tahlequah

attended many of the committee sessions and reported on the proceedings to Charles Eldred, a member of the board. When William F. Rasmus, a Tahlequah storekeeper, told the senators of what he considered the cattlemen's corruptive influence on the Cherokees, Lyons remarked that his testimony was "a tissue of misrepresentations from beginning to end." But Augustus Ivey, the man largely responsible for the hearings, was Lyons's favorite target. The attorney had discovered through diligent investigation that Ivey was guilty of forgery and pickpocketing in Vinita and Tahlequah, so he characterized Ivey and the other critics of the lease as a "little crowd of snakes." [7]

7. John F. Lyons to Charles H. Eldred, July 29, 1885, File: Cherokee Strip Live Stock Association (Section X), IAD, OHS. See also Testimony of William F. Rasmus, January 19, 1885, Senate Report 1278, op. cit., Part I, pp. 189–92. John F. Lyons to Charles H. Eldred, December 1, 1885, File: Cherokee Strip Live Stock Association (Section X), IAD, OHS.

Lyons's comments may have comforted the Outlet cattlemen who paid his salary, but his opinions carried little weight with the members of Dawes's committee. Nevertheless, the hearings resulted in a victory, albeit a slight one, for the association. The senators completed their investigation during the summer of 1885, and their published report appeared a year later, on June 4, 1886. Although charges of bribery were not substantiated, one author has stated, "There is ample reason to believe that the Senate investigation committee did not arrive at the whole truth and that a large sum was really expended in bribing members of the Cherokee National Council to vote for this lease." He cites no evidence to support the statement. The association escaped congressional action for the time being, but the hearings had given its enemies a wealth of material for future attacks on its occupancy of Cherokee land.[8]

8. Senate Report 1278, op. cit., Part I, p. i. Edward Everett Dale, *The Range Cattle Industry: Ranching on the Great Plains from 1865 to 1925*, p. 140n.

Trouble began anew when association agents attempted to obtain an extension of the Outlet lease. The lease signed by cattlemen of the association and the Cherokees in 1883 was scheduled to expire in 1888. In 1886 the directors began their plans for its renewal. By October, Lyons could report that "the out-look is encouraging." Bushyhead favored renewal and promised Lyons that he would "do everything in his power to procure the extension." The attorney spent the month in Tahlequah enlisting support for the association and preparing for the session of the National Council that would assemble on November 1.[9]

An unexpected turn of events disrupted Lyons's plans. On November 9, two men, John Bissill and J. W. Wallace, representing an unnamed syndicate, wrote to Bushyhead and offered to buy the Cherokee Outlet for three dollars per acre. News of the proposition reached members of the association—probably through Lyons—and caused no small amount of consternation. Coupled with a difference of opinion among the cattlemen on the details of the renewal plan, the syndicate's offer threatened to dash the association's hopes for obtaining an extension of the lease in 1886.[10]

A bill for renewal of the lease was defeated in the council on November 26. Under the circumstances, Lyons concluded that he should "let the matter rest where it was until next session." Then, after the National Council adjourned, Lyons learned from Bushyhead that the syndicate's proposition was little more than an attempt to defraud the Indians. By that time, however, it was too late to salvage extension of the lease, and Lyons was forced to wait until 1887.

9. John F. Lyons to Charles H. Eldred, October 5, 1886, File: Cherokee Strip Live Stock Association (Section X), IAD, OHS. John F. Lyons to Charles H. Eldred, October 20, 1886, ibid.

10. John Bissill and J. W. Wallace to Hon. D. W. Bushyhead, November 9, 1886 (copy), File: Cherokee Strip Live Stock Association (Section X), IAD, OHS. John L. McAtee to Charles H. Eldred, November 16, 1886, ibid.

Winter brought inactivity in the matter, and the attorney could write "Every thing is very quiet and business of all Kinds seems very dull." He would crave that respite in the spring.[11]

11. John F. Lyons to Charles H. Eldred, December 4, 1886, File: Cherokee Strip Live Stock Association (Section X), IAD, OHS. John F. Lyons to Charles H. Eldred, December 20, 1886, ibid.

Robert L. Owen succeeded John Tufts as United States Indian Agent at Muskogee's Union Agency in 1886. In late November, he received word that some members of the cattlemen's association in Tahlequah were plying Indians with liquor to secure a favorable vote for their renewal plan. The agent went at once to investigate.

Owen arrived in Tahlequah on the evening of November 26, only hours after the defeat of the bill for extension of the lease. He interviewed several Cherokees and by the next afternoon had obtained affidavits concerning the details of ranchers' activities during the council's session. According to his information, Lyons, Eldred, and Thomas Hutton had arrived in Tahlequah on November 1 and had rented the finest room in the National Hotel. The proprietress, Mrs. Eliza Alberty, who was Bushyhead's sister, told Owen that she had seen "a 2-gallon demijohn, three 2-gallon jugs, and one 1-gallon jug, and forty quart-bottles" of whiskey in that room during the three weeks of the council's session. She recalled observing a number of drunken Indians there at various times and commented that "the disgraceful tramping into this room all night long, and the bad odor it was bringing upon the house" had led her to tell the cattlemen "it would have to stop or they would have to leave her house." Her story, Owen noted, was "corroborated by numbers of other people."[12]

12. Robert L. Owen to Hon. J. D. C. Atkins, April 29, 1887, U. S., Congress, Senate, 50th Cong., 2d sess., 1888, Executive Document 136, p. 3.

In addition, Augustus Ivey informed the Union agent that repre-

The National Hotel, Tahlequah

sentatives of the association paid him $500, ostensibly to settle an "old claim" against Outlet cattlemen. Shortly thereafter, they asked him to support the association's renewal bill. Ivey said the money was paid "really to silence his opposition," but in any case, he kept it.[13]

13. Senate Executive Document 136, op. cit., p. 3.

Owen returned to Muskogee and spent the winter studying the documents he had obtained. Not until spring did he report the situation to his superiors. On April 29, 1887, he mailed the affidavits to Commissioner Atkins and asked for his "immediate consideration" and advice. The National Council was to reconvene on May 9, and Owen expected the cattlemen to renew their efforts to extend their lease. Despite the incriminating evidence he presented against the Cherokee Strip Live Stock Association, the agent told Atkins, "I am of the opinion that it would be best that these lands should be released [*sic*] in whole or in part to the men who now occupy it, provided they are willing to give its fair market value." [14]

14. Senate Executive Document 136, op. cit., p. 4.

Atkins sent Owen's letter to Secretary of the Interior L. Q. C. Lamar, noting that Lyons, Eldred, and Hutton had "made themselves liable to exclusion from the Indian Territory, and to a criminal prosecution in the United States courts" for violating federal statutes "relating to the liquor traffic with Indians." Lamar referred the matter to United States Attorney General Augustus H. Garland for investigation. Garland, in turn, referred it to the United States attorney for the Western District of Arkansas. Lamar then advised Atkins of the pending investigation and said Owen "should be instructed that if he is satisfied the statements as to violations of the law . . . are true and well founded" to eject Lyons, Eldred, and Hutton "from that portion of the Indian Territory within the jurisdiction of his agency."

Robert L. Owen, U. S. Indian agent, Muskogee

There is nothing to indicate that Owen ever removed the men or that Garland ever completed the investigation. The association re-opened negotiations with the Cherokee Nation in the fall of 1887.[15]

15. J. D. C. Atkins to the Secretary of the Interior, June 11, 1887, Senate Executive Document 136, p. 2. L. Q. C. Lamar to the Attorney-General, June 13, 1887, ibid., p. 6 and note. L. Q. C. Lamar to the Commissioner of Indian Affairs, June 13, 1887, ibid., pp. 5–6. John H. Oberly to the Secretary of the Interior, February 26, 1889, ibid., pp. 1–2.

Political strife plagued the Cherokee Nation late in 1887. Rival factions struggling for control of the tribal government narrowly avoided armed confrontation. In December, Joel B. Mayes defeated Dennis Bushyhead in a tumultuous election for principal chief that came dangerously close to bringing bloodshed to the streets of Tahlequah. Bushyhead's defeat was a blow to the association's scheme to extend the lease. Unquestionably, Bushyhead had been the cattlemen's friend; while Mayes was certainly not antagonistic to their interests, he had precisely defined ideas on the disposition that should be made of the land beyond the Arkansas.[16]

16. Morris L. Wardell, A *Political History of the Cherokee Nation*, 1838–1907, pp. 343–44.

On January 25, 1888, Mayes presented to the National Council his views on renewing the Outlet lease. He suggested that the land be leased by competitive bidding for a period of five years. Bids could be invited through advertisements, which would run for three months in newspapers in Arkansas, Missouri, Kansas, Colorado, Texas, and New Mexico. Mayes acknowledged that association ranchers had "dealt fairly and honorably in their obligations to the Cherokee people, and punctual in their payments" of the semiannual rental fee. "Therefore," he concluded, "their interests should be strictly guarded and protected, and should they wish to reoccupy said land, preference should be given them, should no higher bid than theirs be offered."[17]

17. J. B. Mayes to the Hon. National Council, January 25, 1888, File: Cherokee—Strip (Tahlequah), 1888, IAD, OHS.

Joel B. Mayes, Principal Chief of Cherokee Nation, 1887

The National Council ignored Mayes's suggestions. The association's representatives now offered to pay $125,000 per year for a five-year extension of the lease, and a bill to that effect was passed by the Cherokee senate and was sent to the Principal Chief for his approval. Mayes vetoed it, on the grounds that the lease did not result from competitive bidding. "I have reliable information," he told the council, "that similar grazing privileges in the adjoining states and territories are worth from four to twenty cents per acre." The association's offer amounted to less than two cents per acre. Mayes reiterated his belief that the council should advertise for bids on grazing in the Outlet. "When you have done this," he said, "I am sure you have done your full duty to your country and people, and will thereby realize a full compensation for the use of those lands." [18]

18. J. B. Mayes to the Hon. Senate, February 1, 1888, File: Cherokee—Strip (Tahlequah), 1888, IAD, OHS.

The council still refused to act on Mayes's recommendations. In June, a committee from the council's upper house asked the Principal Chief if he had received any bids from cattlemen interested in leasing Outlet range. Mayes remarked that the senators were "certainly aware of the fact that no one has been authorized to receive such bids." He indicated, however, that several ranchers were prepared to bid whenever the Cherokee Nation was ready to consider their offers. He acknowledged having received correspondence from cattlemen on the subject but would divulge no further information because, he said, it would "amount to nothing unless the council authorize the receiving of bids." [19]

19. J. B. Mayes to Hon. Lacy Hawkins, June 27, 1888, File: Cherokee—Strip (Tahlequah), 1888, IAD, OHS.

In July the association increased its offer for the extension of the lease to $150,000 per year. A bill allowing the lease to the association was passed by the senate in an extra session, but again Mayes spurned *the legislation for renewal. Others would pay more for the grazing privilege, he told the council in a veto message. Then he released the information he had been reluctant to reveal only weeks earlier. The North and West Texas Live Stock Company of Dallas, the Chief announced, had offered the Cherokee Nation $160,000 per year for a five-year lease of the Outlet; also, Patrick Henry and D. J. Miller, the Texans who had attempted to lease the range in 1880 for $185,000, were prepared to pay $175,000 for the concession. "I must respectfully recommend," he wrote, "that you proceed immediately to offer to the highest and best Bidder the grazing privilege. . . . I am sure under the present Circumstances that this course would be to the best interest of the Cherokee people." Within a week the Cherokee Strip Live Stock Association raised its offer to $175,000, and the senate passed another extension bill. Once more, Mayes exercised his veto.[20]*

Two considerations governed the Principal Chief's decision. First, the association had not shown any willingness to deposit money with the Cherokee Nation in advance or to produce "a certified check that the money was safe in some National Bank." Mayes believed that "in no instance would the Cherokee Nation be safe to deal with any Company, except by payment of the Money in advance." In a business situation, he said, "the simple promise of any Company will not do." Other cattle companies interested in the Outlet had offered more than promises.

Mayes's second objection to the extension bill was that it did not secure the greatest possible revenue for the Cherokee Nation. When the association raised its bid to $175,000, Henry and Miller increased theirs to $185,000. Still, the council accepted the association offer,

20. J. B. Mayes to the Honorable Senate and Council (envelope dated July 9, 1888), File: Cherokee—Strip (Tahlequah), 1888, IAD, OHS. John B. Wilson to the Cherokee Legislature, July 9, 1888, and Patrick Henry, D. J. Miller, et al. to the National Council of the Cherokee Nation, n.d., ibid. J. B. Mayes to the Honorable the National Council, July 14, 1888, ibid.

and that action, said Mayes, had "the appearance of unfair dealing." In fairness to the Texans, he told the council, "I am sure it would be showing good faith on your part to award this Company this grazing privilege." Five days later, when the council had taken no action, Mayes announced that he was "thoroughly satisfied that it is useless to waste any more of our public fund in trying to arrive at a point where the highest bidder can get this franchise"; he then adjourned the extra session.[21]

21. J. B. Mayes to the Honorable National Council, July 14 and July 19, 1888, File: Cherokee—Strip (Tahlequah), 1888, IAD, OHS.

Two years of lobbying for an extension of the lease had brought nothing to the cattlemen of the association. The agreement was due to expire on October 1, 1888, and in mid-September, realizing that a new lease might not be obtained until the next year, cattlemen began preparing to turn over to the Cherokee Nation fences, corrals, buildings, and other improvements. They declared the structures to be Cherokee property and resolved to surrender them on the demand of any authorized Cherokee agent.[22]

22. John A. Blair to Hon. J. B. Mays [*sic*], September 19, 1888, File: Cherokee—Strip (Tahlequah), 1888, IAD, OHS.

On September 28, only two days before the lease expired, Secretary of the Interior William F. Vilas announced in a letter to Mayes that the federal government would "recognize no lease or agreement for the possession, occupancy or use of any of the lands of the Cherokee Outlet." Should the current lease be extended or a new one signed, he wrote, it

> will be without the authority or consent of this Government thereto, will be subject to cancellation, and any use or occupation by any lessee or any person under such lessee subject to instant termination, by this

> Department at any time whenever such action shall be for any reason deemed proper by the President or this Department, and will be subject to any legislation whatever . . . which Congress may enact affecting that portion of the Cherokee country.

The Commissioner of Indian Affairs informed Robert Owen of Vilas's decision, and the agent prepared notices incorporating the text of the Secretary's letter for distribution among cattlemen.[23]

Mayes carefully studied the Interior Department's directive and on October 10 wrote to Vilas, asking for an explanation of his action. The Secretary replied that the "apparent probability" of a new lease or an extension of the old one had prompted the announcement "to protect the rights of the United States, whatever they are." Still, he wrote, "I am so far from desiring to trench upon the rights of the Cherokee Nation that I strongly wish to see all their rights fully protected." He stated that "no further action appears necessary" and assured Mayes that he intended no threat to the Cherokees' sovereignty. The assurance may have been small consolation to the Principal Chief, but it was all he had.[24]

Vilas's pronouncement had little effect on the aspirations of the Cherokee Strip Live Stock Association. Outlet cattlemen still wanted to obtain a lease, and all that troubled them was a rumor that some members of the National Council would try to pass a measure calling for the sale of the land beyond the Arkansas, presumably to the federal government. Quick action was required if cattlemen were to protect their interests, and leading ranchers in the association soon formulated a plan they thought would appeal to the Cherokees.[25]

Perhaps thinking that Mayes's opposition to their earlier schemes

23. W. F. Vilas to Hon. J. B. Mayes, September 28, 1888, File: Cherokee—Strip (Tahlequah), 1888, IAD, OHS. R. L. Owen to Hon. J. B. Mays [*sic*], October 1, 1888, ibid.

24. W. F. Vilas to Hon. J. B. Mayes, October 23, 1888, File: Cherokee—Strip (Tahlequah), 1888, IAD, OHS.

25. John F. Lyons to Charles H. Eldred, October 23, 1888, File: Cherokee Strip Live Stock Association (Section X), IAD, OHS.

for extension reflected a deep-seated antagonism to the Cherokee Strip Live Stock Association, the ranchers formed a new organization to bid for the land beyond the Arkansas. Chartered in Kansas on October 31, the South Western Grazers Association differed from the older corporation in name only; included among its directors were ten members of the Cherokee Strip Live Stock Association. On November 1 Andrew Drumm, president of the new association, submitted to the Principal Chief a bid of $200,000 per year for a five-year lease on the Outlet and a total of $3,750,000 for a fifteen-year concession.[26]

For three weeks the Cherokees' leaders contemplated the new offer. In the meantime, Cherokee Treasurer Robert Ross announced that, in lieu of a lease arrangement, he had begun collecting taxes under the law of 1879 from ranchers using the Outlet. While the council and the Chief debated, Ross collected $43, 750 in advance for three months' grazing.[27]

Late in November, some members of the National Council tried to pass, over Mayes's veto, the bill accepting the Cherokee Strip Live Stock Association's $175,000 bid. Drumm learned of the attempt and withdrew the higher bid submitted by the South Western Grazers Association. But he acted without the blessing of all Outlet cattlemen. L. P. Williamson of Williamson, Blair, and Company, which subleased pasture from the Cherokee Strip Live Stock Association, was a director of the new organization. Speaking for other directors who perhaps sought to supplant established association leaders like Drumm and Eldred, Williamson renewed the offer of $200,000 for a five-year lease and increased the bid for a fifteen-year agreement to $4.5 million.

The effort in the National Council to override his veto and the

26. Compare Charter of the South Western Grazers Association [October 31, 1888] and List of Lessees on the Cherokee Strip Live Stock Association with post office Addresses [September 7, 1888], both in File: Cherokee—Strip (Tahlequah), 1888, IAD, OHS. A. Drumm and John A. Blair to the Hon. J. B. Mayes, November 1, 1888, ibid.

27. Robert B. Ross to the National Council, November 20, 1888, File: Cherokee—Strip (Tahlequah), 1888, IAD, OHS.

confusion caused by the conflicting bids of the two associations angered Mayes. He finally lost all patience with the Outlet ranchers and charged that the Cherokee Strip Live Stock Association was "attempting to use money to corrupt not only our legislative body, but even the Executive Department of this nation in order to carry out . . . schemes to swindle the Cherokee people out of their just rights." He urged the council to consider carefully its actions "in the name of justice to the Cherokee people." But on December 3, 1888, the council passed a bill leasing the Outlet to the Cherokee Strip Live Stock Association for five years at an annual rate of $200,000. However reluctantly, Mayes approved the measure on December 4, and the renewal fight was over.[28]

28. L. P. Williamson to Hon. J. B. Mayes, n.d., File: Cherokee—Strip (Tahlequah), 1888, IAD, OHS. J. B. Mayes to the Hon. the National Council, November 22, 1888, ibid. Senate Bill No. 37, Act Leasing The Strip Lands [December 1–4, 1888], (copy), ibid.

The cattlemen in the Outlet had narrowly averted disaster to their grazing lease during the years from 1884 to 1888. The investigations conducted by the Senate Committee on Indian Affairs, while failing to prove the Cherokee Strip Live Stock Association guilty of any wrongdoing, succeeded in focusing the attention of federal agencies on the ranchers' lease. The opposition to cattlemen's occupancy of Cherokee land that had begun in the Department of the Interior shifted to the halls of Congress, as evidenced in Secretary Vilas's pronouncement in September, 1888. Owen's revelations about the association's activities in Tahlequah had compromised the stockmen's position, but not until Vilas announced that future leases would not be recognized by the federal government did Washington take a firm stand on the question of use of the Outlet. Perhaps too many agencies and branches of the federal bureaucracy were involved in the matter to permit formation of a coherent, if not consistent, government policy before 1888. Nevertheless, the response of the federal government to the problems in the Outlet conformed to the government's attitude on the larger issue of Indian affairs.

The federal government, in its rulings at this point, was expressing a change in official attitudes toward the Indian and his rights to hold property. Severalty, which emerged from the concept that ownership of private property was an essential mark of civilization and a gift to be bestowed upon the Indian, had achieved the status of law in 1887. Although allotments of Cherokee land were not to be made for nearly a decade, severalty had an immediate impact on the tribe, insofar as it was symptomatic of the government's desire to free the Indian to "enter the market place as an individual entrepreneur." In this context, Vilas's statement acquires new significance. The right to lease land, denied to the Cherokees as a nation, was later granted in severalty, and for many of the Indians it proved to be the only practical means of obtaining revenue from their allotments. In 1888, however, the federal government's action was simply another challenge to the Cherokees' national sovereignty. For their part, the members of the Cherokee Strip Live Stock Association, who had struggled through two investigations and surmounted Mayes's three vetoes to acquire an extension of their base in the Outlet, were not bothered by such matters. They had more immediate problems.[29]

29. William Appleman Williams, *The Contours of American History*, p. 322.

VII. AN ERA PASSES

The nimbus of anomaly that encircled the Cherokee Strip Live Stock Association faded in the glare of the market place. There, cattle produced in the Outlet were subject to the same fluctuations of price that alternately pleased and plagued cattlemen everywhere in the West. Despite their unique position, the ranchers in the Outlet could not escape the disaster that befell the Western range cattle industry after the winter of 1886–1887.

In their zeal to reap profits from the beef bonanza, ranchmen were universally guilty of overstocking their ranges. The winter of 1886–1887 struck hard, particularly on the Northern Plains, and cattle died by the thousands. E. C. Abbott, Granville Stuart's son-in-law, described conditions in Montana:

> The cattle drifted down on all the rivers, and untold thousands went down the air holes. . . . They would walk out on the ice, and the ones behind would push the front ones in. The cowpunchers worked like slaves to move them back in the hills. . . . Think of riding all day in a blinding snowstorm, the temperature fifty and sixty below zero, and no dinner. . . . The horses' feet were cut and bleeding from the heavy crust, and the cattle had the hair and hide wore off their legs to the knees and backs. It was surely hell to see big four-year-old steers just able to stagger along. It was the same all over Wyoming, Montana, and Colorado, western Nebraska, and western Kansas.[1]

1. E. C. Abbott ("Teddy Blue") and Helena Huntington Smith, *We Pointed Them North: Recollections of a Cowpuncher*, p. 176.

The winter was perhaps less severe in the Indian Territory, but the problems of ranchers in the Outlet were compounded by a sudden increase in the region's population of predators. In an effort to curb their losses of calves, the Cherokee Strip Live Stock Association offered a bounty of $1.50 for coyote pelts and $20 for gray wolves.[2]

2. Evan G. Barnard, *A Rider of the Cherokee Strip*, p. 117.

In terms of the dollar value of their cattle, most Western ranchers recovered quickly from the winter's devastation. A succession of wet summers and mild winters created ideal conditions for stockmen on the Northern Plains, but ranchers in the Outlet were less fortunate. Drought struck the area during the summer of 1888, and intense heat scorched grass and blinded cattle. Cattlemen in the Outlet used teams of horses hitched to scrapers in the sandy river bottoms in their efforts to uncover sources of water for their stock. Nevertheless, market prices remained high. The sign belied the times, however, because the bonanza era of Western ranching had passed. Despite a favorable market, as Ernest Staples Osgood has shown,

> the old confidence in the range was gone. Never again would cattlemen dare to take the chances that had been regarded as part of the business in the earlier day. Those who still remained in the business found the margin of profit so small that a winter loss that had been but an average one in the old days would now prove ruinous. The range no longer appeared a safe basis for the industry.[3]

3. Barnard, *Rider of the Cherokee Strip*, p. 126; John Clay, *My Life on the Range*, p. 188. John T. Schlebecker, *Cattle Raising on the Plains, 1900–1961*, p. 6. Ernest Staples Osgood, *The Day of the Cattleman*, p. 224.

Against this economic setting, Outlet ranchers had sought and obtained an extension of their lease from the Cherokee Nation.

The federal government renewed its challenge to the association's occupancy of the Outlet in 1889. An Indian appropriation

Small herd watering in Salt Fork in the 1880s. From ***Indian Territory: A Frontier Photographic Record by W. S. Prettyman,*** selected and edited by Robert E. Cunningham. Copyright 1957 by the University of Oklahoma Press.

act passed by Congress on March 2 contained a provision for the appointment of a commission to negotiate with the Cherokees for the sale of their land west of the 96th meridian. The government's offer was to be $1.25 per acre, or, deducting a sum "chargeable against said lands by direction of certain acts of Congress," $7,489,718.73 for 6,574,486.55 acres of land. The commission began its work, and the federal government proceeded with plans to open the Outlet to settlement.[4]

On October 26 Secretary of the Interior John W. Noble wrote to Lucius Fairchild, chairman of the Cherokee Commission, which was then meeting in Tahlequah, to announce that the government would soon issue an order requiring the association to remove all cattle from the Outlet by June 1, 1890. News of the letter quickly reached cattlemen, and in November members of the association addressed a memorial to President Benjamin Harrison summarizing the history of their occupancy of Cherokee land. Recalling that the eviction of stockmen from the Cheyenne–Arapaho reservation by presidential decree in 1885 had struck a blow to regional cattle interests "by means of which . . . every head of stock in America, on farms and on the range, was depressed in price, and from which those interests have never recovered," they urged the President to consider their economic circumstances. Enforcement of the June deadline, coming as it did at the beginning of the market season, would ruin regional livestock businesses and damage the fragile structure of cattle prices throughout the nation. The quarter of a million beeves pastured in the Outlet could not, cattlemen claimed, be disposed of "without demoralizing and breaking the markets of the country." [5]

4. The activities of the commission are described in Berlin B. Chapman, "How the Cherokee Acquired and Disposed of the Outlet: Part Three—The Fairchild Failure," *The Chronicles of Oklahoma,* 15:3 (September 1937), 291–321. David H. Jerome, Alfred M. Wilson, and Warren G. Sayre to the President, January 9, 1892, U. S., Congress, Senate, 52d Cong., 1st sess., 1891, Executive Document 56, V, pp. 12–13.

5. Donald J. Berthrong, "Cattlemen on the Cheyenne–Arapaho Reservation, 1883–1885," *Arizona and the West,* 13:1 (Spring 1971), 5–32. "A Memorial to the President of the United States from the Members of the Cherokee Strip Live Stock Association," typescript in File: Cherokee Strip Live Stock Association (Section X), IAD, OHS.

Perhaps thinking that their memorial had little or no impact on either the President or the agencies of government charged with acquiring and opening Indian land, the association's directors sought assurance from the Cherokees in January, 1890, that their occupancy of the land beyond the Arkansas would be protected by the tribe under the extended lease. Chief Mayes replied that his government would do what it could to guard the association's interests. "Of course," he wrote,

> the Cherokees have not the warriors to withstand the United States soldiers, that day is passed and gone. Our ancestors fought the battle for our soil, and had to succumb to a superior force. We think, as human beings, living in the land of liberty and free speech, under a Government which proposes to take care of the oppressed, and give justice to all man-kind, that we will be allowed to own and use the soil we bought from that great Government, and we will rely on the law to protect us, and ask the President to use his troops to protect us in the possession of this soil, and by so doing will protect your lease of the grazing privilege.[6]

Within a week Mayes discovered that his faith in federal justice had been misplaced. On February 17 President Harrison issued a proclamation forbidding the further introduction of cattle into the Outlet and ordering the removal by October 1 of all livestock currently there. The four-month postponement of the deadline first mentioned by Secretary Noble was perhaps an indication that the association's plea for time to dispose of its cattle had not fallen on totally deaf ears.[7]

In March, Mayes traveled to Washington to determine what bearing the presidential proclamation had on the Cherokees' rights

6. E. M. Hewins, A. G. Evans, et al. to Hon. Joel B. Mayes, January 24, 1890, File: Cherokee —Strip (Tahlequah), 1890–1891, IAD, OHS. J. B. Mayes to E. M. Hewins, A. G. Evans, et al., February 11, 1890, File: Cherokee Strip Live Stock Association (Section X), ibid.

7. "By the President of the United States of America, a Proclamation," printed circular in File: Cherokee—Strip (Tahlequah), 1890–1891, IAD, OHS.

in the Outlet. Would the Indians, he asked Harrison and Noble, "be prohibited from bringing in stock of their own, for the purposes of pasturage?" Because federal intervention ended occupancy by the association, the tribe acquired under provisions of the lease extension more than one hundred separate pastures fenced with barbed wire worth $250,000. Mayes was reluctant to lose the benefit of those improvements.[8]

The Principal Chief received an answer on March 29. Commissioner of Indian Affairs T. J. Morgan served notice to all cattlemen, "whether white men or Indians," that the October deadline would be enforced.[9]

If Mayes suspected that Harrison's order was, as Dale suggests, "a political move . . . against the Cherokees to force a cession" of the Outlet—the consequence of removing revenue-producing tenants—he found no comfort in the cattlemen's response to it. The Cherokee Strip Live Stock Association's secretary, John A. Blair, sent Morgan a copy of a resolution to the effect that cattlemen would respect the presidential proclamation and "assist the Military in ferreting out and removing" parties that brought new stock into the Outlet. Presumably, they might aid in the eviction of Cherokee ranchers. And if that were not enough to cause consternation in Tahlequah, Charles Eldred then wrote to Mayes asking for "the protection necessary to a peaceable and profitable enjoyment of the grazing privileges" guaranteed to the association by the Cherokee Nation.[10]

The tensions between Cherokees and cattlemen came to a breaking point in August, when Mayes demanded payment of the association's semiannual rental fee, which was a month past due. The

8. J. B. Mayes to the President of the United States and the Secretary of the Interior, March 14, 1890, File: Cherokee—Strip (Tahlequah), 1890–1891, IAD, OHS.

9. "Notice concerning Alleged Cattle Leases on Indian Lands in the Indian Territory," printed circular in File: Cherokee—Strip (Tahlequah), 1890–1891, IAD, OHS.

10. Edward Everett Dale, *Cow Country*, p. 208. John A. Blair to Commissioner of Indian Affairs, June 7, 1890, Letters Received, RG 75, RBIA. Charles H. Eldred to His Honorable J. B. Mays [*sic*], June 7, 1890, File: Cherokee Strip Live Stock Association (Section X), IAD, OHS.

association's president, E. M. Hewins, replied that the money would be forthcoming when the Cherokee Nation could "show . . . any protection in giving . . . the use [of the Outlet] according to contract." His refusal was polite but firm. The association had no funds other than the dues collected from members, and because cattlemen faced eviction, they were in no mood to pay those dues. "I admire your stand as chief executive of [the] Cherokee Nation," he told Mayes,

> as I know you are trying to defend us in our contract with your people, and only make demands that ought to be just and reasonable, but being antagonized, as your people and our association are, by the U.S. government, makes it very embarassing indeed for both of us. I am in hopes that this matter can be brought into court, and there settled definitely.[11]

The Cherokee Nation's Treasurer Robert B. Ross was eager to accept Hewins' suggestion of filing suit to determine the extent of the association's liability to the tribe. The action of the federal government, he told Mayes, did not justify the association "in any attempt to evade the payment of all or any portion of the same they agreed to pay." Hewins had pledged the association "to take all risk of any disagreement with the Federal Government." If the tribe expected to recover any of the rental fee, Ross believed it had best act quickly. The Principal Chief agreed, and the National Council appointed a committee to select attorneys.[12]

While the Cherokees prepared to prosecute the association, the cattlemen proceeded to evacuate their stock from the Outlet. They secured a sixty-day extension, which set the final date for removal on

11. J. B. Mayes to Hon. E. M. Hewins, August 4, 1890, File: Cherokee Strip Live Stock Association (Section X), IAD, OHS. Semiannual payments of $100,000 were due in January and July of each year. E. M. Hewins to Hon. J. B. Mayes, August 11, 1890, File: Cherokee—Strip (Tahlequah), 1890–1891, IAD, OHS.

12. Robert B. Ross to Hon. J. B. Mayes, November 1, 1890, File: Cherokee—Strip (Tahlequah), 1890–1891, IAD, OHS. Thomas H. Barnes and William M. Cravens to Hon. Samuel Mays [*sic*], November 6, 1890, ibid.

December 1. Despite the additional time, ranchers faced a difficult task. Some sold all the cattle they could, and others attempted to move entire herds to new pasture. The exodus created the chaos that Outlet cattlemen had expected.[13]

The forced sale of beef by ranchers of the association seriously affected market prices in adjacent states and threatened the stability of even the largest cattle companies in the West and Midwest. The Matador Land and Cattle Company, a Scottish joint-stock venture in West Texas, paid no dividends to investors in 1890 because of poor market conditions. Cattle sold by the company in Kansas City and Chicago brought only a third of their 1883 price. Much of the Matador's problem was directly attributable to the eviction of stockmen from the Outlet.[14]

Ranchers who chose to move their herds suffered financial loss and untold inconvenience. The Cattle Ranch and Land Company, an English concern that leased pasture in the Outlet through the association, owned no land and was forced to lease acreage from the Texas Land and Cattle Company in Hemphill County, Texas. Under the circumstances, operating costs were prohibitive, and the company could pay no dividends at the end of the season. As a result of eviction from the Outlet, the company lost more than £15,000 during 1891. The firm of DuBois and Wentworth perhaps fared better financially but nevertheless had to drive its 10,000 cattle onto three separate pastures, including one in the Nez Percé reservation, before finding vacant range in Harper County, Kansas.[15]

Despite the publicity that accompanied President Harrison's proclamation and the cattlemen's subsequent departure from the

13. Public Notice, September 9, 1891, printed circular in File: Cherokee—Strip (Tahlequah), 1890–1891, IAD, OHS.

14. W. M. Pearce, *The Matador Land and Cattle Company*, p. 34*n*; Maurice Frink, W. Turrentine Jackson, and Agnes Wright Spring, *When Grass Was King: Contributions to the Western Range Cattle Industry Study*, p. 284.

15. Frink, Jackson, and Spring, *When Grass Was King*, pp. 292–93. DuBois and Wentworth to Charles H. Eldred, June 10, 1892, File: Cherokee Strip Live Stock Association (Section X), IAD, OHS. For the Army's role in removing cattlemen, see Robert C. Carriker, *Fort Supply, Indian Territory: Frontier Outpost on the Plains*, chap. 8.

Outlet, few outside Washington or Indian Territory understood that the federal government intended to open the land beyond the Arkansas to settlement. Perhaps thinking that the federal agencies' objections to cattle grazing in the Outlet arose from antipathy toward the lease concept, large stockmen offered to buy the land from the Cherokees. On December 6, D. R. Fant of Chicago wired Mayes a bid of $10 million for the Outlet. Two days later, Williamson, Blair, and Company of Kansas City offered $20 million, and on December 9 the Lucas Cattle Company of Colorado Springs bid $30 million. There is nothing to indicate that Mayes seriously entertained any of these overtures; but if the bids accomplished little else, they served to expose, at least to the Cherokees, the niggardliness of the federal government in its offer for the purchase of the Outlet.[16]

Although Harrison's proclamation and the association's refusal to honor the provisions of its lease deprived the Cherokee Nation of substantial revenues, Mayes did not abandon attempts to preserve his tribe's interest in the Outlet. Opon learning that cattle were pastured on Outlet grass after December 1 in violation of the presidential edict, he issued an executive order authorizing L. L. Crutchfield, a Cherokee treasury official in Vinita, to collect a grazing tax from their owners. Such action, he told Crutchfield, appeared to be consistent with federal policy. In the meantime, attorneys working for the tribe prepared their briefs against the association, and within six months the stage was set for an encounter between cattlemen and Cherokees.[17]

16. D. R. Fant to Joel B. Mayes, December 6, 1890; Williamson, Blair, and Co. to Joel B. Mayes, December 8, 1890; Lucas Cattle Co. to Joel B. Mayes [telegrams], File: Cherokee—Strip (Tahlequah), 1890–1891, IAD, OHS.

17. J. B. Mayes to L. L. Crutchfield, June 17, 1891, File: Cherokee—Strip (Tahlequah), 1890–1891, IAD, OHS. The executive order was issued on December 2, 1890. A. R. Strother to John W. Noble, December 5, 1890; C. Brownell to John W. Noble, December 29, 1890, Letters Received, RG 75, RBIA.

Attorneys for the Cherokee Nation filed suit against the Cherokee Strip Live Stock Association in the District Court of Sumner County in Wellington, Kansas, early in the summer of 1891. After Chester I. Long of Medicine Lodge presented the association's brief, the Cherokees' counsel W. W. Schwinn asked for a continuance in order to amend the tribe's petition. Hewins's earlier statement about Outlet ranchers' reluctance to pay their dues proved correct; the association had no funds, and a suit against it could not succeed. After the continuance, tribal attorneys announced their intention to sue cattlemen individually.[18]

The association's directors met in July to discuss the problems arising from the suit. They appointed as treasurer Andrew Drumm, who was by then president of the American National Bank and the Drumm–Flato Commission Company, both of Kansas City. To him fell the responsibility for the association's indebtedness, which was estimated to be approximately $6,000, including $1,500 in legal fees. The directors voted to retain an attorney to defend each member in succession, and Drumm set about trying to raise the necessary cash. "It will cost you much less money to join with us," he told ranchers in a form letter soliciting contributions to the defense fund. But by the first of December, cattlemen had subscribed only $422.50; cash outlays totaled $524.60.[19]

While Drumm wrestled with the association's financial problems, the federal commission to negotiate for the purchase of the Outlet continued to meet in Tahlequah. The Cherokee National Council

18. A. Drumm to Dear Sir, n.d. [form letter], File: Cherokee Strip Live Stock Association (Section X), IAD, OHS.

19. A. J. Drumm to Dear Sir, op. cit.; John W. Nyce, Jr., to Charles H. Eldred, July 14, 1891, File: Cherokee Strip Live Stock Association (Section X), IAD, OHS. Cherokee Strip Live Stock Association, December 1, 1891 [ledger sheet], ibid.

appointed its own commissioners on November 16. Eleven days later, the Indians agreed to sell their land for three dollars per acre. Further negotiations ensued, and on December 19 the conferees reached a final agreement. The Cherokee Nation relinquished its rights to the Outlet for the sum of $8,595,736.12.[20]

The Cherokees' decision to sell the Outlet before litigation with cattlemen had been completed seriously weakened the tribe's position in court. As a result, the association's attorneys became confident of victory. The second lease had never been recognized by officials in Washington, they said, and therefore the association could not be bound by its terms. "Rightfully we do not owe them a dollar," Chicago attorney E. C. Moderwell told Andrew Drumm, "and legally they cannot make us pay them one dollar." Nevertheless, the association sent Thomas Hutton to Tahlequah in February, 1892, to attempt a compromise with the Indians. The association's position was that it wished neither to pay the full $200,000 agreed upon in the lease nor to deprive the Cherokees of a reasonable sum for their use of the Outlet during 1890. Drumm told Charles Eldred, however, that he entertained "no hopes of making a compromise."[21]

The association's directors met at the Midland Hotel in Kansas City on March 16 to discuss the Cherokee suit. In the course of their discussion they agreed to reimburse members who had advanced money for the defense fund after "all just debts" of the association had been paid. At the end of the month, Drumm reported to Eldred that members had contributed an additional $983.90 to the fund and promised $250 more.[22]

An inadequate treasury was not the only problem that confronted

20. E. C. Boudinot, I. A. Scales, et al. to David H. Jerome, Warren G. Sayre, and Alfred M. Wilson, November 27, 1891, Senate Executive Document 56, op. cit., p. 16. An act to ratify and confirm certain articles of agreement by and between the United States and the Cherokee Nation of Indians, in the Indian Territory [December 19, 1891], Senate Executive Document 56, op. cit., p. 19.

21. E. C. Moderwell to Andrew Drumm, February 10, 1892, File: Cherokee Strip Live Stock Association (Section X), IAD, OHS. A. Drumm to Charles H. Eldred, February 19, 1892, ibid.

22. Minutes of the meeting are appended to H. R. Johnson to Charles H. Eldred, March 16, 1892, File: Cherokee Strip Live Stock Association (Section X), IAD, OHS. A. Drumm to C. H. Eldred, March 30, 1892, ibid.

the directors as they attempted to navigate through a legal action that was to become a maze of court actions. Trouble developed with John L. McAtee, one of the cattlemen's attorneys. McAtee's principal virtue, as far as the directors were concerned, was his willingness to work for promises rather than for cash. His moderation in the matter of fees, however, was counterbalanced by his incompetence. "McAtee does not make any headway with the suits he has against members of the association," Drumm told Eldred in April. "J. V. Andrews' suit was postponed. I think McAtee was afraid of the Judge." But incompetent in court or not, the lawyer soon sought remuneration for his services, which Drumm ignored. In August, McAtee revealed the lengths to which he would go to collect his fees.[23]

23. A. Drumm to C. H. Eldred, April 22, 1892, File: Cherokee Strip Live Stock Association (Section X), IAD, OHS.

During the summer of 1892, the Sumner County District Court appointed C. A. Gambrill, a Wellington insurance broker, to act as receiver for the association. The directors had given McAtee promissory notes signed by several association members in lieu of their annual dues payment. He was to file a series of suits to collect the money, but the court instructed him to turn the notes over to Gambrill, pending the outcome of the litigation with the Cherokees. The notes were the association's only assets, and Gambrill quickly saw that they would become the focus of attention from the Cherokees. The tribe had already filed its own suits against many of the signatories, all of which meant more work for Gambrill. He asked McAtee to act for him during the prosecution of the various suits, and the lawyer, seeing an opportunity for profit, agreed.[24]

24. John L. McAtee to Hon. C. J. Harris, August 8, 1892, File: Cherokee—Strip (Tahlequah), 1892, IAD, OHS. C. A. Gambrill to Hon. J. C. Harris [*sic*], August 11, 1894, File: Cherokee—Strip (Tahlequah), 1894, IAD, OHS.

On August 8, McAtee wrote to C. J. Harris, Mayes's successor as Cherokee Principal Chief, outlining his arrangement with Gambrill.

The receiver had no money for court costs, and, wrote McAtee, "with the prospect of the proceeds going to the Cherokee Nation, the Association will not advance these necessary funds further." The lawyer believed that Harris should "relieve the situation, and enable us to go forward with the work by appropriating such a sum as seems requisite to get on with the litigation and collection of the notes." The Principal Chief must have been amazed by McAtee's letter: The attorney for the defense was soliciting funds to aid in prosecutions detrimental to his clients. There is no evidence that Drumm and the directors of the association ever learned of McAtee's letter, but even without that knowledge, they were to have more than sufficient reason in succeeding months to regret their choice of attorneys.[25]

25. John L. McAtee to Hon. C. J. Harris, August 8, 1892, op. cit.

On October 24, Drumm received a letter from McAtee asking for money to pay court costs in a suit involving two members of the association. Apparently the lawyer had obtained nothing from the Cherokees. The association's treasury was momentarily depleted, so Drumm supplied the cash from his own account. "I have informed McAtee," he announced in a letter to Eldred, "that I am getting ***very tired puting my hands in my pockets*** and ***paying out money for the Association,*** with a pore show of geting it back." Unable to contain his rage on the subject of the attorney, Drumm wrote Eldred a second letter on the same day, stating that he wanted McAtee to estimate his total bill for services rendered to the association. Considering the lawyer's poor courtroom performance, the treasurer told Eldred, he had advised McAtee "to get through with one suit" before beginning another.[26]

26. A. Drumm to C. H. Eldred, October 24, 1892, File: Cherokee Strip Live Stock Association (Section X), IAD, OHS. A. Drumm to C. H. Eldred, October 24, 1892, ibid.

Problems arising from court costs and legal fees continued to plague the cattlemen. Loans secured to meet the association's obligations came due, and the directors were forced to use their own money to repay them, thereby depleting their personal bank accounts; on October 29, Eldred's bank balance showed a credit of two dollars. The continuing optimism of their legal consultants must have been small consolation to the men who bore the financial burden for the association.[27]

Drumm's sparring with McAtee began anew early in 1893. Writing to Eldred on January 17, he commented,

> I suppose the next suite we will have will be with John L. McAtee. He claims he must be paid for his work and expences done for the Association. He has been beaten in three cases recently and I do not think he will collect one dollar from the delinquent members of the Association.

In an attempt to replenish the association's treasury, Drumm drafted another form letter to cattlemen. His effort brought only $325 and more promises.[28]

Two years of work as treasurer for an organization devoid of capital resources took their toll of Andrew Drumm's patience. Upon learning that a verdict would soon be rendered on the Cherokee suit in Sumner County District Court, Drumm wrote to Eldred to urge dissolution of the association as soon as possible after the decision. "I am tired of the annoyance," he said.[29]

On March 24, 1893, the court dismissed the Cherokee suit on the grounds that the association's lease was in violation of federal law and was thus not binding on either party. W. W. Schwinn, the Cherokees' attorney, quickly planned an appeal before the Kansas Supreme

27. A. Drumm to C. H. Eldred, October 29, 1892, File: Cherokee Strip Live Stock Association (Section X), IAD, OHS. G. R. Peck to C. H. Eldred, December 29, 1892, ibid. Peck was general solicitor for the Atchison, Topeka and Santa Fe Railroad. Chester Long sought his opinion on the Cherokee suit.

28. A. Drumm to C. H. Eldred, January 17, 1893, File: Cherokee Strip Live Stock Association (Section X), IAD, OHS. A. Drumm to S. H. Foss, March 2, 1893, and A. Drumm to C. H. Eldred, March 18, 1893, ibid.

29. A. Drumm to C. H. Eldred, March 20, 1893, File: Cherokee Strip Live Stock Association (Section X), IAD, OHS.

Court. The tribe, he said, had not received a fair hearing. "I believe that the decision of the court would have been different," he told Chief Harris,

> if the Strip had been actually opened for settlement and settled before the decision was rendered. The term of office of the Judge . . . expires next January . . . and he is very anxious to be his own successor. There are such a large number of boomers settled in this County, waiting for the opening of the Strip, that it would have been political suicide for him to have decided in favor of the Cherokee Nation, if these boomers stay here long enough to vote next fall. . . . I feel satisfied that these things had a great bearing on the mind of the Judge, although he may not have been conscious of it.

Seventeen months later, Harris learned from C. A. Gambrill that the court's guiding consideration had been that "Indians could not vote in Sumner County and Cattle men could." [30]

Four days after the decision, Eldred, as the association's acting president, notified the directors of a meeting to be held on April 4 in Kansas City. Topics heading the agenda were John L. McAtee and the Cherokee Nation. Cattlemen knew that the tribe might appeal, and in that event they would need a well-planned defense. Gambrill had returned the association's papers and account books soon after the court proceedings, but McAtee had absconded with the promissory notes. By refusing to surrender the association's only tangible assets, the lawyer jeopardized the cattlemen's position. Without the notes, the organization could not raise enough money to finance an adequate defense in the event that Cherokees appealed the district court's decision, nor could it pay the Indians if that decision were reversed by a higher court. The situation dictated decisive action.[31]

30. W. W. Schwinn to Hon. C. J. Harris, March 28, 1893, File: Cherokee—Strip (Tahlequah), 1893, IAD, OHS. C. A. Gambrill to Hon. J. C. Harris [*sic*], August 11, 1894, op. cit.

31. Charles H. Eldred to E. M. Hewins, March 28, 1893, File: Cherokee Strip Live Stock Association (Section X), IAD, OHS.

The directors agreed to negotiate with McAtee for the return of the notes. The lawyer offered to surrender them for $2,500 and the fees owed him by the association. If the cattlemen preferred, he told Eldred, a settlement could be made for $1,500 and fees. In that event, he would "jointly engage with any one in the suits" arising from the notes. Presumably, McAtee would then share in whatever sum the association recovered from members who were delinquent in their dues. Eldred could not meet McAtee's terms. That, commented Drumm, was unfortunate. The treasurer urged legal action against McAtee to obtain the notes. "If we can . . . turn them over to Chester I. Long," he told Eldred,

> I believe he can collect them. I do not think McAtee an honest honorable man. He is working for McAtee and not the Association. . . . I know some members of the Association are bitterly opposed to paying McAtee $1.00 without being compelled to do so.[32]

Chester Long agreed that some action was necessary. McAtee was still filing suits against cattlemen for their dues to the association, but because the directors would not pay his fees, he was not prosecuting the delinquent members. The cases were being terminated by default, thereby inflicting substantial losses on the association. But while Long expertly defined the implications of McAtee's action, he could suggest no remedy.[33]

Eldred resolved the cattlemen's dilemma on August 12. Acting on the directors' authorization, he transferred "all the right, title and interest of the Cherokee Strip Live Stock Association in and to the notes, claims, accounts and leases to sub-lessees" to Drumm, H. W. Cresswell, and A. J. Snider, "including all accounts, demands and

32. Propositions of compromise made by J. L. McAtee, June 2, 1893, File: Cherokee Strip Live Stock Association (Section X), IAD, OHS. A. Drumm to C. H. Eldred, June 12, 1893, ibid.

33. Chester I. Long to C. H. Eldred, June 12, 1893, File: Cherokee Strip Live Stock Association (Section X), IAD, OHS.

claims . . . arising under the leases and contracts made with sublessees since October 1st., 1888." In return, the three men agreed to hold the association "harmless from all claims, demands, or accounts" made against it by members or by John McAtee and "to defend any suits . . . and . . . pay all judgments, if any, rendered against the Association." A month later, on September 16, the federal government opened the Outlet to settlement, and 100,000 homeseekers swept onto the land once occupied by vast herds of cattle.[34]

If Eldred's transfer dissolved the structure of the Cherokee Strip Live Stock Association, the Outlet's opening to settlers removed all hope of its rebuilding. The organization's name continued to appear on a succession of court dockets, much to the cost of those who had assumed its obligations, until finally it was obliterated by interminable litigation. The Cherokee Nation appealed its case before the Kansas Supreme Court, only to have the lower court's decision upheld; Gambrill threatened to file suit against the tribe for the $1,000 awarded to him by the district court for his services as receiver in the original case; and John L. McAtee, who never obtained his fees from the association, offered to hire attorneys to carry the Cherokees' suit to the Supreme Court of the United States. But the questions that arose after September, 1893, were moot. The Outlet no longer existed, nor did the Cherokee Strip Live Stock Association.[35]

34. Bill of Sale [August 12, 1893], File: Cherokee Strip Live Stock Association (Section X), IAD, OHS.

35. A. Drumm to Charles H. Eldred, September 9, 1895, File: Cherokee Strip Live Stock Association (Section X), IAD, OHS. James Lawrence to Hon. C. J. Harris, February 13, 1894, File: Cherokee—Strip (Tahlequah), 1894, IAD, OHS; Tom George to Hon. C. J. Harris, February 23, 1898, File: Cherokee—Strip (Tahlequah), 1898, IAD, OHS.

Waiting for the signal opening the Cherokee Strip, 1893

VIII. AN ASSESSMENT

Despite the brevity of its existence, the Cherokee Strip Live Stock Association achieved historical significance unsurpassed by surviving regional cattlemen's organizations. Some groups, like the Wyoming Stock Growers' Association, asserted powerful influence in territorial and state politics in the West. Others, like the Texas and Southwestern Cattle Raisers Association, entered a new period of growth and prosperity that extended well into the twentieth century. But the activities of the Cherokee Strip Live Stock Association touched on issues transcending particular regional interest, whether political or economic, and dealt with the larger questions of federal land and Indian policies and the place of the Indian in the national economy. The association—unique among stockmen's organizations—derived no importance from its relationship to the Western range cattle industry, perhaps the most dramatic difference of all.[1]

It is impossible to examine the association in its broadest context without considering the assessments of Edward Everett Dale. In his work, Dale tried consistently to place the history of the cattle industry in a Turnerian mold. "Here," he said of the association, "is to be seen an excellent example of the ability of the American pioneer to organize in a region without law or courts extralegal institutions that seemed to function with surprising efficiency and afford adequate protection to

1. Agnes Wright Spring, *Seventy Years: A Panoramic History of the Wyoming Stock Growers Association*; W. Turrentine Jackson, "The Wyoming Stock Growers' Association: Political Power in Wyoming Territory, 1873–1890," *Mississippi Valley Historical Review*, 33:4 (March 1947), 571–94; Maurice Frink, *Cow Country Cavalcade: Eighty Years of the Wyoming Stock Growers Association*; *The Texas and Southwestern Cattle Raisers Association: Its History, Purpose and Present Day Activities*, [Fort Worth, 1949], p. 4. The parochial interests typical of regional associations are discussed in Mody C. Boatright, "The Myth of Frontier Individualism," *Southwestern Social Science Quarterly*, 22 (June 1949), 14–32, and J. Orin Oliphant, *On the Cattle Ranges of the Oregon Country*, pp. 245–49.

extensive economic interests." But the example of any other cattlemen's organization might better have served his purpose. The unusual circumstances that surrounded the association's organization and operations disqualify it from such a discussion. It was perhaps extralegal in the government's view, but it had a basis in Cherokee law that was recognized, albeit tacitly and intermittently, by several federal agencies, including the Department of the Interior and the War Department. Moreover, it functioned in a geographical region ringed by federal courts, and the success of the homeseekers who invaded Indian land bore boisterous witness to the persuasiveness of those institutions. As for the cattlemen, the courts of Kansas and Arkansas, not the machinery of the association, answered their needs. And the romantic notion derived from Turner of cattleman-as-individualist is disproved not so much by the existence of a cooperative organization as by the ranchers' strong reliance on the efficacy of due process of law in preference to recourse to direct, personal action, even during times of extreme provocation.[2]

2. Edward Everett Dale, *Cow Country*, p. 210.

The federal agencies' interest in grazing arrangements in the Outlet affected the futures of both cattlemen and Indians. Having prompted formation of the association, this interest led eventually to the eviction of cattlemen and the destruction of Cherokee sovereignty. By evicting ranchers from the Outlet and purchasing the land for settlement, the government denied Cherokees the right to conduct their own affairs. Undoubtedly, political considerations motivated federal action, but its consequences were largely economic. "To say," Dale nevertheless has argued, "that the land was taken from the Indian and given over to white settlement is only nominally correct. The Indian as an economic factor was negligible. What really happened was that the land was taken from the ranchman and given to the farmer." But the situation was more complex.[3]

3. Dale, *Cow Country*, p. 211.

Between 1870 and 1893, the question of Cherokee sovereignty was inseparably linked with the Outlet. The extent to which Cherokees could manage it as revenue-producing property became a matter of overriding importance during the administrations of Principal Chiefs Bushyhead and Mayes. Instructions given by the National Council to Cherokee delegates in Washington and correspondence concerning the Chilocco incident contain ample evidence of the significance the tribe attached to control of the Outlet as a manifestation of sovereignty. In that sense, the problems of sovereignty and its maintenance were economic. Given the degree to which Cherokee cooperation was necessary for the effective use of the Outlet by cattlemen or the federal government, the Indian was by no means a negligible factor. The association leased the land with the consent of the National Council and the Principal Chief, and despite its continuing concern with corporate image, it feared the Cherokees' displeasure far more than it did senatorial investigations or bureaucratic intervention.[4]

4. William W. Savage, Jr., "Leasing the Cherokee Outlet: An Analysis of Indian Reaction, 1884–1885," *The Chronicles of Oklahoma*, 46:3 (Autumn 1968), 292.

The sale of the Cherokee Outlet, more than the eviction of ranchers in 1890, marked the passing of the cattleman's last frontier. No ceremony marked its demise, nor was it heralded by the sound and fury that had accompanied the end of the open range a decade earlier. It succumbed later and more quietly, perhaps, than other frontiers to the movement that Turner called "the march of civilization westward," but its passing was no less significant, if only because it announced the death of Cherokee sovereignty.

The Run of '93

The Outlet had been an area in which Indians and white men had achieved a remarkable degree of economic cooperation and had formed, however briefly, a symbiotic relationship. Pressure brought by homeseekers and their advocates in the East provided the final impetus for the federal government's assault on barriers to the land beyond the Arkansas, but the philosophy that undergirded the government's action had origins that antedated the Homestead Act or any other nineteenth-century land legislation. Henry Knox's concept of private property as an instrument with which to civilize the Indian and the Jeffersonian notion that the strength of the Republic lay in the hands of the yeoman farmer were ideas implanted in the collective mind of the federal bureaucracy during the early stages of its development. They emerged full-blown and embodied in government policy decades later. Add to these the Dawes Act, with all its implications, and federal acquisition of the Outlet is explained. The government had determined that the Indian, through severalty, could not function as an entrepreneur in concert with his people. The principle, once established, had only to be applied. Tribal sovereignty succumbed with tribal land, and in December, 1891, the Cherokees lost more than the Outlet. By comparison, the cattlemen lost nothing of consequence.[5]

Litigation with the Cherokee Nation forced the association to remain in existence long after it had outlived its usefulness to cattlemen. It was a body without members, abandoned by ranchers, not, one must conclude, because they feared adverse court judgments, but because they faced the immediate problem of survival in a competitive economy. The Outlet was no longer available, therefore the association was no longer necessary. The ranchers who had belonged to it

5. This is suggested, if not stated, by William T. Hagan, *American Indians*, pp. 43–44, 53–55. The persistence of federal espousal of the yeoman ideal is discussed in William L. Bowers, "Country-Life Reform, 1900–1920: A Neglected Aspect of Progressive Era History," *Agricultural History*, 45:3 (July 1971), 211–22.

were businessmen who labored in a world dominated, not by barbed wire, but by ledger books. They turned to new partnerships, trusts, and corporations; they became bankers, company directors, investors, politicians. This is not to say that they abandoned the cattle industry, but for many of them the Outlet venture had been but a single aspect of their entrepreneurial activity; other enterprises in other places awaited their attention. It remained for their former employees, the men who had ridden the ranges of Indian Territory, to form a sentimental attachment to the land and the organization, to construct a memorial to a way of life that was no more.[6]

6. Evan G. Barnard, A *Rider of the Cherokee Strip*, p. 219.

If ranchers, as businessmen with varied interests, escaped unscathed from the government's action, those whose livelihoods depended on the cattle economy did not. The departure of cattlemen from the Outlet necessitated a radical economic reorientation in the urban centers of southern Kansas. Merchants who once catered to the needs of cowboys sought new customers among homesteaders traveling to the Outlet. Hotels, boardinghouses, and even saloons prepared for a less affluent, less free-spending clientele. Newspapers that once resisted the idea of a farm economy found new publishers, adopted new policies, or died. In a large sense, ranchers left behind a social and economic vacuum in the Kansas cattle towns; despite the adjustments that townspeople made, for many of them the opening of the Outlet dealt a numbing blow to community aspirations. Nevertheless, institutions remained that facilitated the transition. Among these were the courts; the law that prevailed in the day of the cattleman served the farmer equally well.[7]

7. William W. Savage, Jr., "Newspapers and Local History: A Critique of Robert R. Dykstra's *The Cattle Towns*," *Journal of the West*, 10:3 (July 1971), 572–77.

Marker at corner of quarter-section claim. From ***Indian Territory: A Frontier Photographic Record by W. S. Prettyman,*** selected and edited by Robert E. Cunningham. Copyright 1957 by the University of Oklahoma Press.

Finally, the history of the Cherokee Strip Live Stock Association provides a trenchant footnote to the role of the federal bureaucracy in the American West. Recent assessments present inspired arguments in behalf of bureaucracy as a positive force in Western social, political, and economic development. But the problem here is initially one of definition. Accepting Max Weber's dictums on the efficiency of bureaucratic structures, some historians have blithely catalogued what they consider to be the beneficial results of governmental involvement in Western affairs. There exists, however, in the literature of sister disciplines in the social sciences a considerable body of evidence that bureaucratic systems are inefficient, insensitive, and ineffective. The history of federal Indian policy overwhelmingly confirms that view, and the study of the Cherokee Strip Live Stock Association brings it into sharp focus.[8]

Federal insensitivity to Westerners' needs had a long tradition. The problem is not exclusively American; it is common to any government concerned with the administration of a hinterland. If the activities of a government and its various agencies are said to be efficient or even beneficial because they accomplish a specific end—in this nation, the expansion of Anglo-American civilization—that is one measure; if the same government then ignores the requirements of a substantial segment of the population in its charge, that is quite another. The association's members, forced from the Outlet to make way for a wave of settlers, were denied access to land in sufficient quantities for the efficient operation of their businesses, and the Cherokees were deprived of a substantial source of revenue. Having benefited one group at the cost of two others, the government promptly

8. See especially Gerald D. Nash, "Bureaucracy and Reform in the West: Notes on the Influence of a Neglected Interest Group," *The Western Historical Quarterly*, 2:3 (July 1971), 295–305. Marshall E. Dimock, "Bureaucracy Self-Examined," *Public Administration Review*, 4 (1944), 197–207, and Michael Crozier, *The Bureaucratic Phenomenon*. The problems involved in formulating a viable definition of the term suitable for wide application are apparent in Barry D. Anderson, "Reactions to a Study of Bureaucracy and Alienation," *Social Forces*, 49:4 (June 1971), 614–21, and Charles M. Bonjean and Michael D. Grimes, "Some Issues in the Study of Bureaucracy and Alienation," ibid., 622–30. For an overview, see Martin Albrow, *Bureaucracy*. The substantiating literature is enormous, but a recent case study is available in Arrell M. Gibson, *The Chickasaws*.

abandoned the settler—who had benefited—to his struggle with the environment. The debate over the efficacy of the Homestead Act is not without some foundation, but the question is not whether a quarter-section was enough land to sustain the farmer in the West. Rather, it should be why the government thrust a predetermined quantity of topsoil upon men who lacked the "technology, experience, and appropriate type of farming" to earn their living. The Indian farmer was no less disadvantaged than the white.[9]

The businessmen who had leased the Cherokee Outlet were economically more flexible than either settlers or Indians. Practical and pragmatic, they, as had their counterparts in the industrial East, adapted to federal policies and bent to the government's will without breaking. They formed the Cherokee Strip Live Stock Association in response to one policy; they dissolved it a decade later to conform to another. In that way they endured, and the bureaucracy was left to resolve the problems it had created.[10]

9. Robert M. Finley, "A Budgeting Approach to the Question of Homestead Size on the Plains," *Agricultural History*, 42:2 (April 1968), 114.

10. Bill Sampson, "Justice for the Cherokees: The Outlet Awards of 1961 and 1972," Master's thesis, University of Tulsa, 1972.

BIBLIOGRAPHY

Manuscript Collections

National Archives, Washington, D. C.
    Bureau of Indian Affairs, Record Group 75. Letters Received, 1881–1907

Oklahoma Historical Society, Oklahoma City
    Indian Archives Division
        Cherokee Nation Records
        Cherokee Strip Live Stock Association File
    Library
        Athey Collection
        Payne Collection

Western History Collections, University of Oklahoma Library, Norman
    Boomer Literature File
    Cherokee Nation Papers
    Phillips Collection

Government Documents and Publications

Annual Report of the Commissioner of the General Land Office for the Year 1885. Washington: U. S. Government Printing Office, 1885.

Annual Report of the Commissioner of Indian Affairs to the Secretary of the Interior for the Year 1881. Washington: U. S. Government Printing Office, 1881.

Annual Report of the Commissioner of Indian Affairs to the Secretary of the Interior for the Year 1884. Washington: U. S. Government Printing Office, 1884.

Kappler, Charles J., comp. and ed. *Indian Affairs: Laws and Treaties.* Washington: U. S. Government Printing Office, 1904.

U. S. *Congressional Record.* 49th Cong., 1st sess., 1885. XVI.

U. S. House of Representatives. 51st Cong., 2d sess., 1890. Report 3768.

U. S. Senate. 41st Cong., 2d sess., 1870. Report 225.

———. 48th Cong., 1st sess., 1883. Executive Document 54, IV.

———. 48th Cong., 2d sess., 1884. Executive Document 17, I.

———. 48th Cong., 2d sess., 1884. Executive Document 19.

———. 48th Cong., 2d sess., 1884. Executive Document 50, II.

———. 49th Cong., 1st sess., 1885. Report 1278, Parts I and II, VIII.

———. 50th Cong., 2d sess., 1888. Executive Document 136.

———. 52d Cong., 1st sess., 1891. Executive Document 56.

Newspapers

Caldwell Commercial (Caldwell, Kansas). 1880–1883.

Caldwell Post (Caldwell, Kansas). 1880–1883.

The Cherokee Advocate (Tahlequah). 1883–1885.

Oklahoma War Chief (various locations). 1883–1885.

Bibliography

Published Diaries and Memoirs

Abbott, E. C. ("Teddy Blue"), and Helena Huntington Smith. **We Pointed Them North: Recollections of a Cowpuncher.** New ed. Norman: University of Oklahoma Press, 1955.

Barnard, Evan G. **A Rider of the Cherokee Strip.** Boston: Houghton Mifflin Company, 1936.

Clay, John. **My Life on the Range.** New ed. Norman: University of Oklahoma Press, 1962.

Fessler, Julian, ed. "Captain Nathan Boone's Journal." **The Chronicles of Oklahoma,** 7:1 (March 1929), 58–105.

Spaulding, George F., ed. **On the Western Tour with Washington Irving: The Journal and Letters of Count de Pourtalès.** Norman: University of Oklahoma Press, 1968.

Books and Pamphlets

Archer, Sellers G., and Clarence E. Bunch. **The American Grass Book: A Manual of Pasture and Range Practices.** Norman: University of Oklahoma Press, 1953.

Albrow, Martin. **Bureaucracy.** London: Praeger Publishers, Inc. 1970.

Bearss, Ed, and Arrell M. Gibson. **Fort Smith: Little Gibraltar on the Arkansas.** Norman: University of Oklahoma Press, 1969.

Carriker, Robert C. **Fort Supply, Indian Territory: Frontier Outpost on the Plains.** Norman: University of Oklahoma Press, 1970.

Crozier, Michael. **The Bureaucratic Phenomenon.** London: Tavistock, 1964.

Dale, Edward Everett. **Cow Country.** New ed. Norman: University of Oklahoma Press, 1965.

———. **The Range Cattle Industry: Ranching on the Great Plains from 1865 to 1925.** New ed. Norman: University of Oklahoma Press, 1960.

Easton, Robert. **Max Brand: The Big "Westerner".** Norman: University of Oklahoma Press, 1970.

Foreman, Grant. **The Five Civilized Tribes.** Norman: University of Oklahoma Press, 1934.

———. **Indians & Pioneers: The Story of the American Southwest Before 1830.** Rev. ed. Norman: University of Oklahoma Press, 1936.

———. **Indian Removal: The Emigration of the Five Civilized Tribes.** New ed. Norman: University of Oklahoma Press, 1953.

Frantz, Joe B., and Julian Ernest Choate, Jr. **The American Cowboy: The Myth & the Reality.** Norman: University of Oklahoma Press, 1955.

Frink, Maurice. **Cow Country Cavalcade: Eighty Years of the Wyoming Stock Growers Association.** Denver: The Old West Publishing Company, 1954.

———, W. Turrentine Jackson, and Agnes Wright Spring. **When Grass Was King: Contributions to the Western Range Cattle Industry Study.** Boulder: University of Colorado Press, 1956.

Gard, Wayne. **The Chisholm Trail.** Norman: University of Oklahoma Press, 1954.

Gibson, Arrell M. **Oklahoma: A History of Five Centuries.** Norman: Harlow Publishing Corporation, 1965.

———. **The Chickasaws.** Norman: University of Oklahoma Press, 1971.

Gittinger, Roy. **The Formation of the State of Oklahoma, 1803–1906.** University of California Publications in History, VI. Berkeley: University of California Press, 1917.

Hagan, William T. **American Indians.** Chicago: The University of Chicago Press, 1961.

Hofstadter, Richard, and Michael Wallace, eds. ***American Violence: A Documentary History.*** New York: Vintage Books, 1971.

Leckie, William H. ***The Buffalo Soldiers: A Narrative of the Negro Cavalry in the West.*** Norman: University of Oklahoma Press, 1967.

McCallum, Henry D., and Frances T. McCallum. ***The Wire That Fenced the West.*** Norman: University of Oklahoma Press, 1965.

McCoy, Joseph G. ***Historic Sketches of the Cattle Trade of the West and Southwest.*** Kansas City, Mo.: Ramsey, Millett & Hudson, 1874.

Morris, John W., and Edwin C. McReynolds. ***Historical Atlas of Oklahoma.*** Norman: University of Oklahoma Press, 1965.

Nye, Russel. ***The Unembarrassed Muse: The Popular Arts in America.*** New York: The Dial Press, 1970.

Oliphant, J. Orin. ***On the Cattle Ranges of the Oregon Country.*** Seattle: University of Washington Press, 1968.

Osgood, Ernest Staples. ***The Day of the Cattleman.*** New ed. Minneapolis: University of Minnesota Press, 1954.

Pearce, W. M. ***The Matador Land and Cattle Company.*** Norman: University of Oklahoma Press, 1964.

Pelzer, Louis. ***Marches of the Dragoons in the Mississippi Valley.*** Iowa City: The State Historical Society of Iowa, [1917].

Rister, Carl Coke. ***Land Hunger: David L. Payne and the Oklahoma Boomers.*** Norman: University of Oklahoma Press, 1942.

Schlebecker, John T. ***Cattle Raising on the Plains, 1900–1961.*** Lincoln: University of Nebraska Press, 1963.

Spring, Agnes Wright. ***Seventy Years: A Panoramic History of the Wyoming Stock Growers Association.*** Cheyenne: Wyoming Stock Growers Association, 1942.

The Texas and Southwestern Cattle Raisers Association: Its History, Purpose and Present Day Activities. N.p., n.d. [Fort Worth, 1949].

Wardell, Morris L. **A Political History of the Cherokee Nation, 1838–1907.** Norman: University of Oklahoma Press, 1938.

Webb, Walter Prescott. **The Great Plains.** Boston: Ginn and Company, 1931.

Williams, William Appleman. **The Contours of American History.** New ed. Chicago: Quadrangle Books, 1966.

Articles

Anderson, Barry D. "Reaction to a Study of Bureaucracy and Alienation." **Social Forces,** 49:4 (June 1971), 614–21.

Berthrong, Donald J. "Cattlemen on the Cheyenne–Arapaho Reservation, 1883–1885." **Arizona and the West,** 13:1 (Spring 1971), 5–32.

Boatright, Mody C. "The Myth of Frontier Individualism." **Southwestern Social Science Quarterly,** 22 (June 1941), 14–32.

Bonjean, Charles M., and Michael D. Grimes. "Some Issues in the Study of Bureaucracy and Alienation." **Social Forces,** 49:4 (June 1971), 622–30.

Bowers, William L. "Country-Life Reform, 1900–1920: A Neglected Aspect of Progressive Era History." **Agricultural History,** 45:3 (July 1971), 211–22.

Chapman, Berlin B. "How the Cherokees Acquired and Disposed of the Outlet: Part Three—The Fairchild Failure." **The Chronicles of Oklahoma,** 15:3 (September 1937), 291–321.

———. "How the Cherokees Acquired the Outlet." **The Chronicles of Oklahoma,** 15:1 (March 1937), 30–49.

Dimock, Marshall E. "Bureaucracy Self-Examined." ***Public Administration Review,*** 4 (1944), 197–207.

Finley, Robert M. "A Budgeting Approach to the Question of Homestead Size on the Plains." ***Agricultural History,*** 42:2 (April 1968), 109–14.

Frantz, Joe B. "The Frontier Tradition: An Invitation to Violence." In Hugh Davis Graham and Ted Robert Gurr, eds., ***Violence in America: Historical and Comparative Perspectives,*** pp. 127–54. New York, Toronto, and London: Bantam Books, 1969.

Gibson, Arrell M. "Ranching on the Southern Great Plains." ***Journal of the West,*** 6:1 (January 1967), 135–53.

Jackson, W. Turrentine. "The Wyoming Stock Growers' Association. Political Power in Wyoming Territory, 1873–1890." ***Mississippi Valley Historical Review,*** 33:4 (March 1947), 571–94.

Nash, Gerald D. "Bureaucracy and Reform in the West: Notes on the Influence of a Neglected Interest Group." ***The Western Historical Quarterly,*** 2:3 (July 1971), 295–305.

Savage, William W., Jr. "Barbed Wire and Bureaucracy: The Formation of the Cherokee Strip Live Stock Association." ***Journal of the West,*** 8:3 (July 1968), 405–14.

———. "Leasing the Cherokee Outlet: An Analysis of Indian Reaction, 1884–1885." ***The Chronicles of Oklahoma,*** 46:3 (Autumn 1968), 285–92.

———. "Newspapers and Local History: A Critique of Robert R. Dykstra's ***The Cattle Towns.*** " ***Journal of the West,*** 10:3 (July 1971), 572–77.

———. "The Rock Falls Raid: An Analysis of the Documentary Evidence." ***The Chronicles of Oklahoma,*** 49:1 (Spring 1971), 75–82.

Turner, Frederick Jackson. "The Significance of the Frontier in American History." American Historical Association, *Annual Report of the American Historical Association for the Year* 1893, 190–227.

Theses

Bradfield, Larry L. "A History of Chilocco Indian School." Master's thesis, University of Oklahoma, 1963.

Sampson, Bill. "Justice for the Cherokees: The Outlet Awards of 1961 and 1972." Master's thesis, University of Tulsa, 1972.

INDEX

A

B

C

D

E

F

G

H

I

J

K

L

M

N

O

P

R

S

T

U

V

W